ENERGY,
CONVENIENT SOLUTIONS

The definitive book on energy for the 21st century

ENERGY, CONVENIENT SOLUTIONS

HOW AMERICANS CAN SOLVE THE ENERGY CRISIS IN TEN YEARS

HOWARD JOHNSON

First edition - an earlier version was published with the name, *A Convenient Solution*
Copyright © 2002 - 2010 by Howard Johnson.
Publication date, August 21, 2010

ISBN: Softcover 978-0-982-91140-2

This book was printed in the United States of America.

This book is available at special discounts for bulk purchases for sales promotions, premiums, fund raising, or educational use. For details, contact:

Special Sales - Senesis Word

FAX: - 904-825-0222

Website: www.Senesisword.com

Email: Senesisword@yahoo.com

978-0-982-91140-2_txt11212

The cover photo: Raw, wild energy on a galactic scale, the striking crab nebula as photographed by NASA using the Hubble telescope.

A nebula starts when a star can no longer support itself, and becomes unstable. The star will eventually blow up, and the dust and chunks of the star will orbit the core. The heat reflects off the chunks, and makes the nebula glow. Rarely, black holes form the same way as a nebula.

M1: The Crab Nebula is the result of a supernova seen in 1054 CE. It is filled with mysterious filaments that are not only tremendously complex, but appear to have less mass than was expelled in the original supernova and a higher speed than expected from a free explosion. The cover image is from a composite photo taken by the Hubble Space Telescope. It is presented in three colors chosen for scientific interest. The Crab Nebula spans about 10 light-years. In the nebula's very center lies a pulsar: a neutron star as massive as the Sun but with only the size of a small town. The Crab Pulsar rotates about 30 times each second.

The Crab Nebula is the expanding remnant of a star's supernova explosion viewed and recorded by Japanese and Chinese astronomers nearly 1,000 years ago in 1054, as did, almost certainly, native people all over the Earth. This composite image was assembled from 24 individual exposures taken with the NASA Hubble Space Telescope's Wide Field and Planetary Camera 2 in October 1999, January 2000, and December 2000. It is one of the largest images taken by Hubble and is the highest resolution image ever made of the entire Crab Nebula.

For more information on this spectacular object, go to:

http://www.seds.org/messier/m/m001.html

Howard Johnson says, "*Technology can provide us with beauty, such as the marvelous Hubble photographs, and solutions to our most vexing problems such as diseases, the energy crisis and the search for understanding, building on past knowledge and probing into the unknown.*

"*Technology can also bring us devastation, terrorism, environmental destruction and even the end of life on our planet. The choice is up to us. We can work together intelligently and build a better tomorrow for everyone or we can turn this green planet into a wasteland?*

"*Yet technology is neither the hero nor the villain in these scenarios. It is the hand of man wielding technology that yields both the creation and the destruction.*"

CONTENTS

SECTION I - Preliminaries

SECTION II - Some General Information

SECTION III - Energy Systems: Old, New, and Future

CONTENTS

CONTENTS

CONTENTS

CONTENTS

CONTENTS

SECTION IV - Politics Rears its Ugly Head

SECTION V - References and Recommendations

CONTENTS

SECTION VI - Appendix

CONTENTS

Dedication

It is with great humility I express my gratitude to all of my family and friends who have endured hearing and reading my technical rumblings about the energy crisis and what should be done about it. They have been tolerant of my passion, my *techno speak*, and of numerous essays on the subject that I asked them to read and critique. Thanks to my sister and brother-in-law, Bobbie and Bob Grimm, for their considerable emotional and financial support of this effort. Without that support, this book would not exist. I especially appreciate and treasure the memory of my late wife, Barbara, who was my editor, proofreader, counselor, and constant support during much of the early time spent on this book even to her last days. These pages reflect much of her effort. She is surely cheering the publication of this work from on high. Thanks also to Daphne Fox whose help and support have been invaluable for the last years of my writing. Among the many who helped me in this endeavor are two others I especially want to thank. They are John O'Renick, who sent me an invaluable critique of an earlier manuscript and Al Kalar who made several excellent suggestions about the layout of the book and some of the content. Their excellent critiques enabled me to refine and improve the book and make it easier to read.

PREFACE

About ten years ago I started writing a book titled simply, SOLUTIONS! In it I proposed practical solutions to many of the knotty problems facing our nation and even the world. I developed solutions to serious problems like drugs, the environment, tax systems, and national security, among others. I described various practical solutions, making each a chapter in the book. Among the many problems examined were two complex and related ones—energy and fuel systems. In 2003, while working on this, I heard an interview on National Public Radio about the substantial promise of the hydrogen fuel cell. The gentleman speaking explained with enthusiasm how it was going to revolutionize the transportation industry with vehicles that only exhausted pure water. It sounded quite promising to one who had worked and done research in the petroleum and energy industries often during the previous fifty years.

On the Internet, I found his website and emailed him about my interest in his work. I asked him if he would provide access to more information about this new technology. He replied quickly thanking me for my interest and providing me with a list of references, articles, and books on the subject, many he had written. I also began looking into the realities of the entire system of which the vehicle and its source of power are but a small part. By the time I had discovered what the whole system would entail: the raw materials, manufacturing processes, distribution, storage, and dispensing of hydrogen, the infrastructure required for such an undertaking, and the new technologies required to create all these interacting systems, it looked to be more than a daunting task. It looked prohibitively expensive. When I emailed him asking about infrastructure costs, he referred me to another member of his *staff* saying they would answer my query. Several unanswered emails later, I received a notice from *his assistant* informing me that the staff member I had emailed left for other employment, for greener pastures I presumed, and that I would soon be hearing from another. Months passed during which time I repeatedly emailed his office without any response. A few months later I received a failure notice from Yahoo. His email was no longer active. So much for the touted *expert* on the hydrogen fuel cell vehicle. Perhaps his government grant ran out and was not renewed.

This piqued my curiosity, heightened my interest, and brought to my attention the growing public concerns about energy and the environment. I began researching energy,

energy systems, fuels, transport, and all the other parts of the complex, interactive systems that comprise energy. Added to what I had learned from my education and years of experience, it became a fascinating store of information—practical data about systems from past, present, and future. I have cataloged much of this information in *Energy, Convenient Solutions* along with my opinions about the forces that will control how we deal with the problems, the motives of those making crucial decisions, and the technologies involved. I'm certain there is much I have missed and much waiting in the wings to be discovered and touted by those who do such things. That's how it is with virtually every item of science and technology. By the time information is published, it has been changed or replaced by a new discovery, system, or use of technology.

The reader may notice some repetition of facts and descriptions. This is because many facts or descriptions fit into several different areas covered in the book. Rather than use cross references that could cause confusion for the reader, many of these usually small parts are simply inserted in the new position within different points of reference. Some simply have more than one place in the orderly progression where they are a necessary fit.

Introduction

This book is about energy, energy systems, energy use, fuels, and fuel use. It describes some history of energy and fuels, their sources, practicality, and uses. It also describes many new and revolutionary materials and systems that could be solutions to the current energy crisis. The best combination of the solutions described could solve our energy crisis in just a few years, a decade at most. The real problem is in enacting these solutions. Implementation will be dependent on varied systems of interacting disciplines, companies, researchers, investors, and governments.

The author recognizes and uses a number of language conventions that are now quite common with which he does not agree. For example, carbon dioxide is quite commonly referred to as a *greenhouse gas*, a serious misnomer. The physical processes by which all gasses, including carbon dioxide, absorb, hold, and radiate heat energy in the atmosphere is completely different from that which holds heat in an actual greenhouse, the *greenhouse effect*. Also and related, *Global Warming* has come to have a very specific meaning that the author finds is far more an emotional belief system than a provable reality. Nevertheless, these terms and others, are used in the text where they convey their now common meanings. Incidently, some individuals are now applying another term to carbon dioxide that is categorically false. That term, applied strictly for political reasons, is *pollutant*. Carbon dioxide is no more an air pollutant than is water, argon, or nitrogen and the term, *pollutant* will not be used to describe carbon dioxide in this book. There are other terms that are similarly incorrect but have crept into the language by common usage.

The many forces that will shape our energy systems

Energy, fuels, and all their associated products and services make for some complex and interacting systems on an immense scale. This rapidly changing, worldwide set of systems is affected by a broad range of factors and circumstances. Some of the key ones include,

1. The state of the world's economy
2. Supply/demand balance
3. World prices of crude oil
4. The politics of nations and organizations that produce and sell crude oil

5.   The politics and power of the oil-importing nations

6.   Supposed global warming and its effects on policy and markets

7.   The *global warming* movement and the power it wields

8.   Profitability of alternate fuels compared with petroleum products

9.   Profitability of various energy use systems

10.   Profitability of various energy generating systems

11.   Government involvement at many levels

12.   Private investment

13.   Public and private research efforts

14.   The news media and even the world of entertainment

There are certainly many more, but to try to list them all would be foolish and counterproductive. Suffice to say that there are enough interacting variables to tax even the expert operators of the most sophisticated super computers. What this means is that significant changes in any of these factors can affect a number of others and not always in predictable ways.

There are many new and old ideas and systems described herein.

The author makes no apologies for favoring some over others. I favor mostly those that seem to be practical, economical and especially speedy. Some of these have come to prominence recently and thus are not covered as thoroughly as others. Things are changing rapidly in this field with new ideas and products appearing almost daily. The rapidly fluctuating price of petroleum triggers many of these changes. It is my belief that an energy shortage or crunch is coming much sooner than most expect. The long-range forecast is for oil prices to spiral upward. The worldwide recession has temporarily halted the rapid rise over the previous few years, but sooner or later the recession will abate and oil prices will continue their long-range rise. Systems described in this book or new ones not yet imagined will eventually replace petroleum because of market forces.

The purpose of this book is to provide information and encouragement for the doers, movers and shakers in our nation—the entrepreneurs. The many energy systems described run from those used for several hundred years to those just discovered and in their infancy. Many of these will fall into disuse or be kept for historical or sentimental usage.

Here's a bit of old news: For all practical purposes, the horse and buggy have left the American scene. Except for the Amish and some nostalgic sight seeing uses, they have disappeared. The Stanley Steamer and the Baker Electric, once quite popular are now found only in museums or in the hands of collectors. The *iron horse* of the plains is but a memory with a rare few still in collections or on sightseeing railroads. A few World War I Sopwith Camels and World War II Japanese Zeros are still flying. Last fall I witnessed a Boeing B17E Flying Fortress fly by while I was walking on a popular Florida beach. It was quite a thrill watching that half a century old legend still flying. How vastly different it is from the modern B2 bomber. This illustrates the increasing speed with which technology advances. Most of what we have today would be as unrecognizable during the era of the B17 as the B17 would have been at the time of the revolutionary war.

As time passes, the evolution of technology accelerates. The sum of scientific knowledge approximately doubles every fifteen years. This has been going on since the time of Copernicus in the fifteenth and early sixteenth centuries, Galileo and Kepler in the sixteenth and early seventeenth centuries and Newton in the seventeenth and early eighteenth centuries. Western science seems to ignore the work of Muslim mathematicians and astronomers who knew that the earth was a sphere and revolved around the sun centuries earlier than Europeans. They, in turn, had learned from Greek and Persian astronomers and mathematicians after translating much knowledge into Arabic from Greek and Persian. These early scientists in turn probably learned much of their knowledge of mathematics and astronomy from the Egyptians.

What about today? With computers to record our work and the Internet to distribute it, new knowledge quickly spans the globe as the sum of knowledge continues its geometric expansion. Not only are we learning new things faster, but new, practical and sometimes serendipitous findings now spread around the world at the speed of light. Information can be distributed instantly, but the actual creation of new items, systems, procedures and processes still requires time and considerable effort to move from raw material to finished products. Most of these fall by the wayside because of unattractive appearance, lack of understanding of their actual value, lack of economic appeal, or even erroneous perceptions. If another item or system is cheaper with the same value or even more expensive but with superior value, that system will prevail. Except within government bureaucracies, profitability is the clue to the economic success of any item or system. The life of even a well-accepted technology can soon be eclipsed by a newer, better or cheaper technology. Witness the evolution in recorded music from the wax cylinder to the brittle 78, the flexible LP and 45, to reel tape, to 8-track tape, to cassette tape, to CD, to DVD, and now micro chip and Ipod. The effective life of each system lasted for a shorter period of time than its predecessor. This is the nature of accelerating technology in the music business as in many other industries.

This book is about the same kind of event happening in the energy industry, a much broader field than music, with many more variations and possibilities. A problem or need arises. Creative minds search for answers, primarily to find ways to make money, a living, even wealth. They present these many answers to the public in ways from simple word-of-mouth contacts to mass media advertising. All things being equal, the highly advertised will always prevail over the word-of-mouth simply because it reaches far more people in a shorter time. By the time widget A gets started by word of mouth, widget B has thousands of orders from its massive advertising.

So it will be with the new systems described in this book. Individuals and companies of many different sizes pursue these proposals or developments, some with patents, some without. I have some distinct opinions of the systems we can and should end up using. Time will provide the reality to my opinions. The same can be said for those interim products needed to move us from total dependence on petroleum to multiple energy systems in the next decade or so. I also have some definite opinions about those I think will not be successful. I freely share these opinions with readers throughout the book.

It makes no difference who or what is blamed for rapidly rising fuel prices, or where they go in the future. It makes no difference what your position is on environmentalism. It makes no difference what the reality is about global warming. It makes no difference how much oil companies are hated or loved. None of these change the fact that we still need alternative fuels and energy sources. They have become an absolute necessity because of diminishing supplies of petroleum.

Use of any fossil fuel will add carbon dioxide to the atmosphere. There are only two known ways to use energy without adding carbon dioxide to the atmosphere. The same two methods will apply whether or not we are forced to survive without petroleum fuels for any reason.

The first and most obvious is energy from fuels derived from plant materials—nonfossil sources. Carbon dioxide created by the burning of these fuels came originally from the atmosphere. Thus, use of fuel produced from plant sources only returns carbon dioxide to the atmosphere that was originally taken from it. These no net carbon dioxide fuels include many common materials: wood, ethanol from corn, methane recovered from landfills, methanol, butanol, DMF and ethanol from plant material fermentation, oils from plant sources including soy beans, palms, and algae, pelletized agricultural waste, and any other form of fuel from recent biological activity.

The second way is, and promises to be, far larger than no net carbon dioxide fuels both today and in the future. It includes all noncombustion processes for generating energy. Those energy systems currently in use include nuclear, river water, solar, wind, tidal water, ocean wave action, and geothermal. Each of these has its own set of challenges, including practical limits, funding, new technologies, environmental impacts, site locations, weather problems, real or imagined dangers, government controls, and concerns of the public.

Any or all of these processes could be used to generate electric power for grid distribution in the *optimal energy economy* of the future as described in these pages. It remains for some nation or organization to take the high road to the cheap, safe, portable, no net carbon dioxide-producing energy that these processes promise. Once in use, the benefits to the economy of any nation that uses it will be unlimited.

Nuclear power, is it passé? In the past, nuclear power has been touted as the best way to produce safe, clean, energy without producing carbon dioxide. Unfortunately, a very slanted and scary movie, *The China Syndrome*, so frightened the American public that the entire nuclear industry was scuttled at tremendous expense and waste. Once more, perception of the American people trumped reality. This false perception was generated by a fictional story. It baffles me that the public believes a completely fictional story over the obvious reality. Hollywood must still be gloating over the destructive power wielded by their movie. It is interesting to note that it was based on an actual nuclear accident, the one at Three Mile Island. The interesting thing about that accident is that **the safety features of the plant worked.** The danger was contained just as the plant was designed to do. Radioactive leakage was far less than the maximum considered safe and the resulting dispersed radiation was barely detectable above normal background radiation. There was never any detectable radiation danger. Fortunately for France and China, they didn't believe or ignored the intended message of the movie, understood the reality, and are now rapidly developing and building nuclear power plants. By the way, nuclear power has been proven the safest of all types of power plants in real terms of human lives lost and bodies injured. I wonder why Hollywood and the media never acknowledge that fact?

Following are some recently released estimates showing the present distribution of the various worldwide energy sources. Also shown are two potential energy sources and how they could stack up for the future.

Hydro electric	*15%*	*2,665 Terawatts*
Nuclear	*15%*	*2,665 Terawatts*
Natural Gas	*20%*	*3,481 Terawatts*
Oil	*7%*	*1,218 Terawatts*
Coal	*40%*	*6,963 Terawatts*
Renewable fuels	*2%*	*348 Terawatts*
Geothermal	*1.25%*	*268 Terawatts*
Wave Potential	*200%*	*34,816 Terawatts*
Geothermal Potential	*1000+%*	*160,000+ Terawatts*

The last two on the list could turn out to be the best in all ways including economic. Geothermal power could be the real winner in an all-out competition given that useable geothermal energy is available in about 60 percent of the area of North America, and similarly throughout the world. This is covered in sections II C and III A 5 on geothermal power. I wonder if Hollywood will mount a new attack on progress with *The China Syndrome II* about a cataclysmic geothermal volcanic explosion. Do not put it past them. Right now in California, several geothermal plants have been supplying power for some time. Though still a tiny part of the overall mix, geothermal power has the most long term potential of any system, including wave energy power.

The marine energy sector is in its infancy compared with all the other energy sources we use today, including geothermal. It's only now starting to gain a lot more attention, and what is more important, the large influx of investment capital it needs to expand. Wave action power generation of the ocean is a recent technology with great promise. Like geothermal, it is already in use for a tiny portion of our electric power. While wave energy is only a possibility to many people, the truth is it is now a practical reality. Several ocean energy companies are not only producing power right now, but they are landing power purchase agreements with the major utilities. No better proof exists that this power generation system is viable than a power purchase agreement. A small Canadian firm that few people even know exists recently picked up a long-term deal with a major utility in California to deliver power to the grid.

Here's what we are facing: Pundits now report that the coming change in energy is certain to be the most drastic and overwhelming disruption the energy markets will ever see. Besides water, there is nothing more critical for the entire world than adequate supplies of cheap energy. We rely upon it for our transportation, our food, our medicine, our clothing, our agriculture. It's the underlying force that keeps the world moving. As we've already begun to see with oil, it is also the one thing that can bring the global community to its knees, if there is not enough of it. So needless to say, an energy resource that is immeasurable and inexpensive is an energy resource that will drive the next evolution of our energy economy. There are not many proven technologies to choose from right now.

Why Petroleum Will not be the Answer

Back in March of 2005 I read a dire prediction about petroleum. It was a confirmation of what I and many in the oil industry have known and studied for as long as 50 years. We have known and predicted the growing, rapid decline in world oil production between the year 2000 and 2025 even that long ago. The March 2005 prediction said we were about to run out of oil. Actually, that is not true. It should have said the discovery and extraction of new crude combined with existing supplies was not keeping up with demand. It correctly reported the price of oil was about to go through the roof. Oil was predicted to reach $80 a barrel within the next two years and go as high as $185 a barrel.

Steve Forbes couldn't resist ridiculing this prediction. He made his own prediction, "In 12 months, you're going to see oil down to $35 to $40 a barrel. It's a huge bubble, I don't know what's going to pop it but eventually it will pop. You cannot go against supply and demand, you cannot go against the fundamentals forever."

The last part of his statement was right on the money. You cannot go against supply and demand forever. That was more than three years ago and now it's reality. Crude oil passed $130 a barrel in May of 2008, and everybody from President Bush to OPEC to the CEOs of Big Oil now say exactly what that prediction was saying in 2005. The world's supply of easy oil is quickly running out. In spite of this, the current economic down turn quickly brought crude oil prices down dramatically. Strange how the recession made Steve Forbes' prediction come true. A quick economic turnaround and oil prices will return to the stratosphere. This pause in rising prices could provide us the time to convert to alternatives, but that is not likely to happen. Besides the human nature to put things off, venture capital required to develop alternative energy has suddenly dried up. The increase in taxes promised by our new government will further inhibit investment.

Little has changed even though all of these seem to have gotten the message: government officials, oil company CEOs, even consultants to the petroleum industry worldwide. They have responded with statements like,

"Growth in global demand for oil is accelerating and the supply is not."

"The era of cheap energy is over, permanently."

"Access to oil and gas can no longer keep up with the demand."

"Prices of all petroleum products are poised to go through the roof."

Then there is my own prediction made early in 2007 in the manuscript for this book of $200 per barrel petroleum and $8 a gallon gasoline for the U.S. in 2010. When I first included that full page prediction, I wrote it as a scare tactic, an attention getter, a way to capture the imagination of the reader. Little did I realize it would be a fairly accurate prediction. It is still there in the middle of Section II of the book along with a new prediction of what will happen with low oil prices. The recession of 2008 brought about a short, two to four year delay of the inevitable.

As the world's oil production slows and the demand for oil rises, the results could be catastrophic. Prices were rising precipitously, not only on oil and oil products, but on virtually every other product or commodity. The first indicators of the looming disaster, rising prices for food and then other items are already evident. Grain prices doubled in the years before the 2008 recession as grain was taken from the food supply to make biofuels. The ripple effect of this switch began creating shortages in poor areas of the world where starvation is a major problem. The world recession of 2008 temporarily reversed these price increases, but by early 2010 they were rising once more. The rapid drop of petroleum prices and the cost of fuel at the pump pleased most Americans. Of course, the job losses and business failures that accompany these dropping prices are not very pleasing. When and if economic stability and economic growth return, oil prices will once again head for the stratosphere. This will only get worse until and unless we develop the new energy systems described herein. The long-range prospects remain for less and less oil at higher and higher prices.

In their edition of May 12, 2008, *The Maine Sentinel* reported, "The modern world needs cheap oil like the human body needs oxygen; remove it, and we could be headed for economic decline, resource wars and social chaos." To me, if cheap oil is like oxygen then even more so is the broader term, cheap energy. Cheap alternative fuels and cheap and plentiful energy are both essential to the health of the world's economies. To prevent monumental economic disasters for the whole world, some individual or group must come up with viable solutions to cheap fuels and energy. Viable energy alternatives are certainly within our grasp. It is vital that we develop these into practical, working systems.

High prices for virtually everything could lead to lower demand, but this could spiral into a very bad depression. In view of the rapidly increasing demand for oil in China followed closely by India and several other nations, economic disaster could be upon us soon and will be the most serious challenge the modern world has ever faced. Hungry and angry people lead to desperate people which in turn can lead to horrible consequences. Should the price of oil and energy continue to escalate it will eventually be priced beyond the ability of ordinary people to pay for it. At this point the economic collapse will be sudden and catastrophic. No developed nation is equipped to handle such a collapse. That's why we must act now—immediately and decisively. Delaying will lead to widespread conflict and even war—war unlike any we have ever seen.

Although most people still believe we have plenty of oil and natural gas and that the prices will soon return to previous levels, others are beginning to realize that is just not true. Left leaning politicians and the talking heads on TV are still saying how we can solve the problem with conservation and new technologies. Reducing our consumption of oil, it will fall back to less than fifty dollars a barrel. That places them firmly among the glue-sniffers. In all seriousness, how can they possibly believe this will happen? This is especially true for the pundits and analysts who regularly appear on television to talk about how improved technology will continue to lower energy costs and bring as much energy to market as we demand. This will force the price of oil back down to $35 a barrel. It will never happen in that way. Market forces will always control the price of oil even as it has dropped the price precipitously because of the deepening recession of 2008. Even if we opened up all the fields in and around our nation to drilling, it would only delay the problem and not for long at that.

Again, remember Steve Forbes' infamous prediction in 2005 that higher oil prices would cause supply to increase and outpace demand. But, according to Matthew Simmons, the world's top oil investment banker and an energy adviser to President George W. Bush, the idea that cheap oil would last forever is a 21st-century myth: "The religion was faith-based, not fact-based! It was an illusion!" At the first Association for the Study of Peak Oil and Gas (ASPO) conference in 2005, Simmons observed that the peak oil problem had started to look like a *theological debate*, and quoted Dr. Herman Franssen, saying, "It is time to leave 'I believe' inside a church." The facts are that our largest oil reservoirs are running out of oil and their production is falling. Most of the world's current oil production is from fields that are past their prime and are now declining. These fields include most of the world's biggest and most productive.

Kuwait's Burgan Oil Field—In an incredible revelation early in May of 2008 it was reported by the Kuwait Oil Company that its Burgan field, the world's second largest oil field, is tapped out and has passed peak output.

Cantarell, The Third Largest Oil Field in the World, Petroleos Mexicanos (Pemex), Mexico's state oil monopoly, expects its production at the Cantarell oil field to slow earlier than previously forecast. Their chief executive said the decline is now expected to average 14% a year starting in 2007 and go down soon after.

Most of the other known reserves of petroleum are in fields that are at least beginning to decline. New fields are getting smaller and harder (read more expensive) to find and bring into production. This has been going on for at least ten or fifteen years. Even the latest oil recovery technologies have had less than dramatic results. Instead of increasing the amount of oil available, these techniques have brought about the more rapid depletion of the existing reserves. The future for cheap oil looks even grimmer as these technologies have hastened the demise of existing oil reserves and reduced the promise of future production. This is already a factor in the rapid rise of the cost of crude.

Add to this, the enormous oil deposits offshore and in Alaska that have been removed from exploration and production almost exclusively by over zealous environmentalists. Then there are those proven fields in our country where the cost of drilling and extraction is between $20 and $30 a barrel. These fields, including one in North Dakota that holds as much as a fifty-year supply of sweet crude, were never tapped when crude could be purchased for $10 a barrel. Now that crude prices have gone so high and it becomes economically feasible to mine, it will take several years to drill, reach, and pump enough of this oil to make any impact. Drilling will take a sizeable investment which comes only from the profits of the oil companies. Should the government, as suggested, increase the taxes on those oil companies, this oil will take just that much longer to be made available. Those politicians and media talking heads never mention that while whipping up public animosity toward Big Oil, do they? They do not want you to know their efforts are the largest contributors to the high prices you must pay for fuel and those efforts are the chief reason we are sending trillions to despotic states that plot our destruction.

Many oil experts both in and outside of the industry correctly predicted the rising prices of crude almost to the dollar as long ago as early 2005. What amplifies the problem is the fact that for every calorie of food consumed in the United States, there were 10 calories of fossil fuel consumed to make the fertilizers, pesticides, and herbicides; fuel to run the machines that plant, tend, harvest, transport, and process the goods; and fuel to deliver them and refrigerate them. That is without considering the energy you use going to and from the stores and then to cook your food. This means that as fuel prices rise, everything that includes a cost of fuel in their mix will rise along with fuel. The extensive use of cheap fossil fuels in food production is what has enabled the world population to multiply by four and a half times in the last century to around 6.7 billion people at the present.

It's quite simple; food is fuel and energy. Food travels an average of 1,300 miles from the farm to the plate in North America, leading critics such as James Howard Kunstler to decry the *3,000-mile Caesar salad* that travels from California's breadbasket, the San Joaquin Valley, to his table in Scranton, Pennsylvania. We need oil for nearly everything we do, and our entire infrastructure is built on the assumption that there will always be lots of it. Serious problems and expensive shortages are no longer coming. They are already here.

"A Saudi oil-output hike would not solve U.S. problems." George Bush 10:04 A.M. May 17, 2008.

U.S. President George W. Bush said that a hike in oil output by Saudi Arabia would not solve American energy problems. "It's not enough, it's something but it doesn't solve our problem," Bush told reporters in Egypt's Red Sea resort of Sharm el-Sheikh. Bush said he was *pleased* with a Saudi decision taken on May 10 to increase its oil production by 300,000 barrels per day in response to customers, but said that he was *also realistic* about what the Americans should do.

"Our problem in America gets solved when we aggressively go for domestic exploration. Our problem in America gets solved if we expand our refining capacity, promote nuclear energy and continue our strategy for the advancing of alternative energies as well as conservation," he said. "It is divided into three comprehensive parts The Crisis in a Barrel, Making Money from the Fossil Fuels That Are Left, and Energy after Oil."

The first two are only band-aids on the problem and merely delay our eventual succumbing to depletion of crude supplies, and not for long. The third is the only option we have and that is what this book is about.

A dangerous reality most politicians, the media, and the public seem to ignore is that the billions of dollars of investment required to power the twofold answer to the energy crises—new oil and alternative energy—must come from oil company profits. Increasing taxes on business will lower this amount substantially and discourage exploration, research and development. Substantial profits of American business are essential to our economic health and to finding solutions to the real energy crisis. The economic explosion of China and other countries will cause the price of crude to keep right on growing past $130 per barrel and heading for $200. Witness the following news report:

China's crude demand is expanding at 11% a year. China has already passed the U.S. as the emitter of the most carbon dioxide in the world and will soon replace the U.S. as the world's biggest oil importer. The growth of India's oil demand is not far behind. These two nations account for a third of humanity. As their breakneck development continues, the energy needs of their factories and construction firms along with those in Brazil, Mexico and other populous emerging markets can only escalate.

Specifically, as these countries get richer, and their citizens can afford more, the number of cars in the world, now around 625 million, is set to double in less than 20 years. Think of the impact of that on global oil demand, seeing as around 70% of current crude output is used to fuel cars.

Quoted from the UK Telegraph, April 2008 (before the economic crisis broke)

But wait just a minute! The imminence of *peak oil* may not be as threatening as we've been warned. In an article in the October 2009 issue of *Scientific American*, author Leonard Maugeri reports on advanced technologies that offer ways to economically extract nearly as much oil known to be underground as has already been delivered. This could extend the actual supply available well into the next century at around current crude prices that fluctuate between $50 and $80 per barrel in 2009 dollars. This means that competitive fuels and energy systems will of necessity need to be in the same or lower range of cost or they will simply not be viable for a very long time. Steve Forbes' price predictions may not have been so far off the mark after all. Such information is certain to frighten away some investors now considering alternative fuels and energy systems. It will cause others to become nervous and cautious about investing their money in new energy. It will also displease the gurus of global warming.

SECTION II - Some General Information

What this Book Is Really About

**Perhaps the best way to explain what this
book is about is to tell what it is NOT about.**

**It is definitely not a hand wringing message
of doom, gloom and contempt for America.**

We have far too many of these messages of doom and gloom given to us daily in the media and by politicians. These vitriolic elitists have nothing good to say or predict about America or Americans. They seem to be doing everything they can to discredit, take away, and destroy all the things most Americans—actually most people in the world—want for themselves and their families. Mostly the availability of everything people want boils down to E-N-E-R-G-Y and what it costs—energy to light our cities and our homes, power our factories, move our vehicles, operate our computers, fly our airplanes, power our medical technology, grow our crops, and build our buildings—energy that does so much for us every day. Of course, fuel is but one part of the energy equation.

Two opposing views of how to manage energy come from differing political viewpoints. One is to utilize the systems proposed in this book to expand energy systems and grow our domestic economy. The other is the way of those who would limit its use, and reduce consumption. Mostly they would use government to enforce stricter and stricter limits, often by levying taxes to artificially raise the price and so reduce use. These people and the power they wield is covered in a later section titled, *Politics Rears its Ugly Head,* starting on page 121 and going through page 159.

It is Not about Solutions in the Distant Future

This book proposes solutions in years, instead of decades, with little infrastructure changes using existing technologies. These solutions are based on total energy systems including creation, storage, distribution, use, power grid stations, fuel manufacture, waste disposal, local power generators, vehicles and vehicle power systems. Not to examine and develop these alternative energy sources is economic suicide.

It Is Not Just about the Growing Demand for Oil

It is interesting to note that the rapidly expanding economies of India, China, and some other third world nations are demanding increasing amounts of petroleum and will continue to do so for years into the immediate future. China is currently on a binge of building power plants and developing sources for petroleum, even near our Gulf Coast. Since the Florida legislature had the wisdom to prohibit American companies from drilling for oil in the Gulf of Mexico near the coast of Florida, our friends, the Chinese, in cooperation with our friends, the Cubans, are now drilling for that oil a few miles off our coast. By using slant drilling techniques, they will be able to extract oil from beneath our continental shelf off Florida and Louisiana. They are not restricted by the safety and environmental rules American companies must abide by, so they can do it the cheap and dirty way. So much for the wisdom of our politicians in protecting our Gulf Coast from oil spills. See *The Realities of the Gulf Oil Disaster* on page 134 and then on page 150 for more information about this major environmental disaster. It seems even with our government controls, disasters still happen.

It Is Not Just about Alternative Fuels

The only real question is, can we convert to alternative fuels fast enough to avert economic disaster? These fuels alone may not provide the solution as they bring about problems of their own, like competition with food. What we actually need is new and more practical energy systems for generating electricity.

It Is Not Just about New Types of Vehicles

The part of the energy use system that the public most responds to and the media most reports about are snazzy new cars. They are also among the last essential parts needed for our overall energy systems. Without a complete operational system to distribute energy from source to vehicle, those cars are merely a useless hunk of unmovable metal and plastic.

It Is Not Just about Reducing Global Warming

There are several overwhelming reasons why we must quickly develop new, innovative energy systems to create and distribute energy. Ideally, these systems will not require fossil fuels or new and expensive infrastructure. Supposed global warming caused by carbon dioxide is the least of these reasons. Even without this consideration, we desperately need an alternative to petroleum products. Thanks in large part to limits imposed by our over zealous and intrusive government, they are becoming more difficult and expensive to find and recover. A sudden, major disruption of the oil supply would wreak havoc with the world economy. It could create a depression that would make the one in the thirties look mild in comparison. This is not an American problem, but a worldwide one.

It's Not about Waiting for a Major Catastrophe

Many of the concepts and systems described are already in existence. We have started to design, build and even use some of these advanced nonfossil fuel systems. This major shift away from petroleum fuels must be made quickly enough to avert the catastrophic economic menace that rising prices for petroleum fuels promise. Those accelerating prices are even now beginning to bring serious economic problems down on the entire world. An adequate solution could probably be found within the systems described in these pages.

It Is Not Just about an Economic Bonanza

Should we develop programs using these systems, the benefits to our nation and the world would be substantial and almost immediate. The optimal energy system would provide far more material benefits than just economic growth and prosperity.

It IS about Preventing Economic Collapse and War

Make no mistake, the real threat of new kinds of war looms larger each day. This tension is fueled by the growing demand for energy from those large nations now experiencing explosive economic growth and demanding more oil as their economies accelerate. The dangerous conflict in The Republic of Georgia was most likely one of these over control of energy. This is compounded by an accelerating food shortage that is possibly even more dangerous than the fuel shortage. As the prosperity of these large nations grows, the demand for fuel and food is far outstripping the supply. The result can be hungry people running amuck in killing frenzies as has happened in much of Africa. Add Islamic fundamentalist terrorists from nations awash in oil money and we have two easily recognizable groups that care nothing about human lives and would not hesitate to snuff out a few hundred million. They would also cheer loudly at the murder of virtually every person in the West. That we find new, nonpetroleum-based systems for energy generation, transport and use is essential to help prevent this from happening. The answer to this can certainly be found by pursuing some of the avenues laid out in these pages. Hopefully, an abundance of cheap energy that does not interfere with the food supply will relieve some of that danger, as well.

It's Not about Words and Emotional Reactions

We always have plenty of that from politicians and the world of entertainment including the media whose stock in trade is the use of words to stir emotions. These voices, frequently of doom and gloom, often falsely condemn many who could be instrumental in solving problems. In fact, they can be causing considerable damage by dividing us and generating discouragement and conflict. They use class envy and contrived figures to entice anger and distrust among the people for the very organizations that are best equipped to solve our many, growing problems.

What we need is positive action—actually many actions by creative people who do much more than talk—and the leadership to help guide and inspire us all. We desperately need people who design and build, the men and women with creative minds and laboring hands who are willing to work hard to provide us both the ideas and the actuality of new energy technology. We need those skilled and hard-working hands that till the soil, build the infrastructure, and operate the computers and machinery, and yes even those who manage and invest. These are what drive the productive engine that has been and hopefully will continue to be America.

Those people are there, now, hard at work trying to solve our problems in the old-fashioned way, American ingenuity and drive. Spurred on by the promise of tremendous rewards if their efforts are successful, those who participate are many, often unknown. The promise of profits—a dirty word to the ignorant and those who would control them—is the fuel that drives the creative human engine that could solve most of our problems if given the chance. It is these free entrepreneurs and investors who will solve the energy crisis if only those posturing and strutting politicians and government officials would stay out of their way.

This book tries to describe the wonders entrepreneurs have created and the ones that will solve our energy problems.

For those who think I am a bit over critical of our government let me say that I appreciate and applaud the effort of those few dedicated public servants who work hard within the burdensome bureaucracy and help our nation. My criticisms of government are of the indolent, make-work leeches in the bureaucracy created by self-serving politicians, and those many self-serving politicians themselves.

I have gained much information from DOE web sites:

http://www.pi.energy.gov/, and

http://www.pi.energy.gov/documents/newecon_appendix.pdf

America Needs a Mission for Energy Independence

That mission is to discover, develop, and implement practical ways to save us—the United States and the world—from the ravages of the fossil fuel dragon. We should do our utmost to make everyone aware of available options for safe, affordable energy generation and use. We should also try to motivate entrepreneurs to pursue the development of as many of these options as are found to be practical, while continuing to look for new and better ones.

It is paramount that we develop realistic solutions to the energy crisis from among the multitude of products and systems that are in use, under development, or even latent ideas in the minds of America's creative genius. We must collect and examine descriptions of fuels and energy systems—past, present, and future—and of many possible and practical ways to replace fossil fuels with renewable fuels or energy systems. It matters not to a driver what powers his vehicle when he presses down on the accelerator pedal. Any power system that provides adequate mobile power economically when that pedal is pressed will satisfy his needs. All of the new systems could replace fossil fuels as the prime energy source for our nation and even the world. In the process, this could lead to a carbon-dioxide-neutral energy system, one that adds no new carbon dioxide to our atmosphere. The options needed are real and practical alternatives to fossil fuels that will replace the use of petroleum and coal-based fuels with renewable, nonpolluting fuels or electrical energy and in the process:

1. build an American energy system that will stop the hemorrhaging of billions of U.S. dollars, mostly to despotic nations that preach our destruction.

2. build an American energy industry that boosts our economy and provides good jobs—real, productive jobs for many Americans.

3. stop the growth of atmospheric carbon dioxide and that possible link to global warming: and accomplish most of this within just the next ten years.

Our total energy system consists of many types of energy systems, sources, fuels and conversions. The requirements of the components of such a workable system should be judged by the following criteria:

1. They should be comparatively inexpensive to use.

2. They should be developed using environmentally sound, sensitive principles.

3. They should be far easier, simpler and less expensive to implement than systems exemplified by the hydrogen fuel cell vehicle.

4. They should be adaptable to our existing infrastructure with minor changes.

5. They should use raw materials we already have or that can be developed here, locally.

6. They should be applicable to existing vehicles with upgrades or conversions.

7. New fuels should be useable with existing IC (Internal Combustion) engines of all types.

8. They should be developed using existing, evolving technology able to be essentially complete within ten years.

9. They should create a system that is a net zero contributor of carbon dioxide to the atmosphere.

10. They should use evolutionary as opposed to revolutionary changes—these will lead to constantly improving, adapting systems driving numerous growing and improving technologies.

11. They should be developed by America-based industry with the many resulting substantial benefits to our nation—social, political, and economic.

While the main thrust of such systems will be to provide new, better, less expensive and less environmentally intrusive systems for energy and transportation, many benefits other than just getting away from fossil fuels accrue. These include direct positive effects on four of the first seven of the *top twenty-two most serious concerns of the American public* as shown in a public survey conducted by MIT and cited below.

Public Perceptions and Concerns

Howard J. Herzog, a principal research engineer at the MIT Laboratory for Energy and the Environment (LFEE); MIT graduate student Thomas E. Curry; and professors David M. Reiner and Stephen Ansolabehere developed a survey including questions about the environment, global warming, and climate-change-mitigation technologies, and the most important issues facing the United States today.[1] The survey in its entirety can be viewed at the following website:

http://sequestration.mit.edu/pdf/LFEE_2005-001_WP.pdf

Questions showed that the environment in general and climate change in particular are not high-priority issues for the public. The environment came out thirteenth on a list of twenty-two possibilities for *the most important issues facing the United States today*. The front-runners on the list were terrorism, health care, and the economy. On a list of ten specific environmental problems, *global warming* came up in sixth place, well behind water pollution, destruction of ecosystems, and toxic waste.

IMPORTANT NOTE: This list represents public perception of the severity of a problem, **not the reality**. It is well known that media attention to a particular problem or situation influences public opinion. Since the survey was taken, and with the growing hype about global warming, that concern now tops the list of environmental concerns, having moved from sixth to first in just a year. That could well be described as the *Chicken Little* effect. Whether or not it is an actual cause for concern is irrelevant. Public perception and the assumption by so many public figures that human created carbon dioxide is causing catastrophic climate change for the worse makes it a real concern to many people. Some realities of our current understanding of climate change including global warming are described in the section on global warming starting on page 161 in this book.

The solutions recommended in this book directly relate to and could be a powerful and positive force toward the following items on the survey, showing their position of importance to U.S. citizens according to the survey taken in 2005. The number two concern, health care, though not directly affected would benefit from the economic growth these solutions would provide.

No. 1 Terrorism
No. 2 Health care
No. 3 The economy
No. 4 Unemployment
No. 7 Federal budget deficit
No. 13 The environment

Changes in Three Years

Public perception has changed considerably since 2005 because of many factors that include the rising cost of petroleum and petroleum products, the rising cost of food along with worldwide shortages, the lack of any significant terrorist attacks on the United States, and the general acceptance as a proven fact that global warming caused by human production of carbon dioxide from fossil fuels represents a real and imminent threat. Add to that the long ago predicted mortgage meltdown and it becomes evident our economy has taken several damaging blows. This latest economic blow was brought about by foolish and even unscrupulous lending practices and speculation that should have been illegal, but were not. Once more our politicians hurry to lock the barn door after the horse has been stolen.

Those who are complaining the loudest are the very ones who orchestrated this crisis. Now they have the gall to ask to be put in charge of the solution. How stupid do they see the public? Are they right? That all of this together has not brought on a collapse of our economy is a testimonial to the strength of that economy. Just how well it continues to grow or decline depends on many interwoven factors. Not the least of these is the result of political actions. The specter of increased taxes and government controls on business in America looms large in the minds of business managers and owners all around the world. People will always react to their perceptions rather than to realities particularly regarding poorly understood phenomena. This is amplified in importance by the media's preponderance to report in detail any bad news or frightening scenario. Add to that those politicians and media personalities who use any possible bad suppositions no matter how insignificant as bludgeons with which to batter any who would dare to disagree with them or their agendas. The effect on the public's perception of virtually anything is powerfully influenced by all the ranting and raving.

A poorly supported yet probably fairly accurate report on the current state of public's perception provides the following new list of related concerns:

No. 1 The economy, especially as it relates to rising food prices, the mortgage meltdown and suddenly lower fuel prices
No. 2 Unemployment
No. 3 Federal budget deficit
No. 4 Terrorism
No. 5 Health care
No. 11 The environment (global warming leads)

This is strange considering that until the sudden economic downturn, the economy continued to expand and unemployment had risen only to around 6%, a normally acceptable level in good economic times. The public's attention has shifted from those figures and in 2008 focused on recession, the mortgage and corporate credit debacle, unemployment, and rising energy and food prices that they see and deal with every day. These every-day realities truly frighten them. With politicians and the media constantly waving the recession flag for more than a year, it is no wonder people are nervous. In fact, deliberate pessimism of the media was probably a powerful force in creating the recession or at least making it worse.

A Monumental Task with Many Obstacles

Even with these substantial benefits bundled into grand plans, the planners must still deal with significant forces. These forces can make a new idea work or relegate it to the ash can of history. Real difficulties and obstacles must be overcome in order for any new system to become a reality no matter how positive and/or effective that system might be. The battle to get the most beneficial systems noticed and made a reality may require more effort than the implementation of the idea or system itself. The process, once begun, may take unexpected twists and turns in moving, sometimes forward and sometimes back, but always in the ultimate direction of success.

Our space program and its goal to *put a man on the moon in ten years* followed just such a wandering path en route to its success. We can expect no less from our efforts to find a new fuel/energy system that has a far more powerful practical and obviously profitable goal. Clearly, President Kennedy's commitment to put a man on the moon in ten years and the follow-up on that commitment was a major force in making it happen. Media hype and glamorization helped garner public support and enthusiasm. That was a government program operated by a government agency implemented mostly by private contractors according to government bid specifications. It was a **process** oriented solution with a single defined goal.

What we need now is leadership that is courageous enough to state a goal such as *convert to new, home based energy systems in ten years* and then work ceaselessly toward achieving that goal. We need leadership that will initiate a system oriented, broad-spectrum approach to solving our growing energy crisis. This is an even greater challenge than putting a man on the moon. It is a serious challenge that could be instrumental in securing our survival. We need this ten-year goal declaration to be well stated and backed by leadership with the vision and dedication to follow it through. The commitment would be to develop new energy systems that will provide American-made renewable fuels or other portable energy systems and will add no more carbon dioxide to the atmosphere and do it within the next ten years.

This commitment is a much broader goal than putting a man on the moon. It has many branching and interconnected avenues that could lead to successful solutions. The key to final success will be found in the development of many areas of research rather than just one or two. These include the best combination of energy sources, means of obtaining that energy, means of moving the energy from source to point of use, and finally the systems of using that energy. A variety of equally effective systems fitting differing needs could be developed by a diverse group of privately funded entrepreneurs and inventors. The result could become a variation on the current theme where we use several types of fuels in different configurations for similar purposes.

Delaying that, as many are now doing by talking about reaching that goal in thirty or fifty years, is a recipe for disaster. We do not have that kind of time to wait. Just run the numbers. Continuing to transfer billions of American dollars offshore for thirty to fifty years will destroy us economically long before we can develop an alternative fuel economy. Even ten years could be too long, but I believe we can handle that. Certainly it would be less disastrous than thirty to fifty years.

What we do not need now are politicians that use class envy, and every negative action they can promote as smoke screens to hide their own obvious and damaging failures. Not just to hide their gross negligence, but to use false factors as the reasons for new and oppressive taxes, and government powers to control commerce and punish those they see fit to punish for any reason. These power-hungry opportunists and their lackeys in the media ridicule and oppose anything proposed or suggested by anyone who is not in their camp.

Here's just one example of the many new energy sources available. This is information about one large oil field from the U.S. Geological Survey. It is the official results of a groundbreaking study released on June 9, 2008. The report confirmed a massive oil reserve in an area the locals have nicknamed *the Bakken*, which stretches across North Dakota, Montana and southeastern Saskatchewan. The study estimates an immense 3.65 billion barrels of undiscovered oil in the Bakken. Compared with the agency's estimate back in 1995, the study reports a 25-fold increase in the amount of oil that can be recovered. The reported mean estimate of 3.65 billion barrels of oil is for undiscovered oil only, and does not include known oil. The total amount of recoverable crude in The Bakken deposit could be as much as 400 billion barrels. Once impossible to extract, this oil has yielded to new horizontal drilling and rock fracturing techniques. The Bakken is now being hailed as the single largest oil find in U.S. history. Experts estimate that this light, sweet (low sulfur) crude will cost Americans about $16 a barrel. Let's hope we can obtain major production from this field before opportunistic obstructionists can figure out a reason to prevent drilling there.

It may be that the current crisis can be diverted by new recovery technology in this field, but hopefully the incentive to produce viable nonfossil fuels and other energy systems will continue. Eventually we will run out of fossil fuel and will need alternatives. The attention given to new energy and fuel systems will undoubtedly involve effort into other seemingly unconnected areas. We are still deriving long-term benefits from technology developed for our space program. It would certainly be the same for any fuel/energy program. It is amazing to discover that so many of our serious problems are interrelated and how finding one solution often leads to another almost totally unrelated solution and so to the demand for another workable system.

Existing Systems

Presently there are at least seven petroleum-based and mined fuels used in a variety of engines and boilers. These are in addition to coal used mostly in power plants. Use of all of these fossil fuels adds carbon dioxide to our atmosphere. There are at least six nonfossil-based fuels currently being used or being considered for use. Most are manufactured from plant materials and add no net carbon dioxide to the atmosphere in use. Some do add carbon dioxide in their process of manufacture. There are a few nonfossil solid fuels, mostly used for heating and cooking. A wide variety of harvesting and manufacturing processes are used to obtain or make these fuels. Some of these manufacturing processes require more energy input than the resulting fuel can produce.

There is also the special case of nuclear fuels that use radioactivity to generate heat to boil liquids that drive turbine generators. Since these do not use combustion, they do not release carbon dioxide to the atmosphere.

The only reason we need fuel is to provide heat energy which we then convert to electricity or mechanical power. There are at least five combustion-based systems in use. The internal combustion piston engine is the most common and the most developed. Turbine engines make up the rest of the internal combustion types. Other sources of power include piston steam engines, turbine steam engines, several types of nuclear reactors, fuel cells, and batteries. All of these power sources turn energy derived from chemical reactions or nuclear fission into electricity or mechanical energy which then powers vehicles, tools, and factories.

There are at least six types of batteries in use, some of which are new and just beyond the development stage. These new technologies will come of age when continuing development of improved technologies lower their costs and improve their safety and efficiency.

Electric motors of many types and sizes, long important in stationary applications and semi-portable tools, are growing in use in vehicles. The fastest growing application of new battery technologies is now battery-powered, cord-free tools and electronic equipment. Application of these new batteries to hybrids, plugin hybrid electric vehicles (PHEVs), and even pure electric vehicles (EVs), is just in the beginning stages.

In the power plant segment of our energy system there are at least eight different sources of energy used to drive the generators that produce our electricity. Each has its own positives and negatives and all can pose serious environmental problems.

All of these parts of our energy system have been described to illustrate how complex it is. Making any major change would be a difficult and arduous task. Even deciding which changes to make—what system to develop—will be difficult. The answer could lie in a very successful technique used mostly in America for a long time, individual entrepreneurship in an unfettered, free-enterprise, business environment.

The Challenge Ahead

There are literally thousands of individuals using their genius to develop new energy technologies motivated by the promise of rewards for themselves and for their organizations. We are not alone in free entrepreneurship. The powers that control China have suddenly realized its value and are now encouraging it. This has created one of the biggest economic turnarounds in the history of nations. Other countries have seen the light for some time and their economies are booming. Even India, the other Asian giant, is beginning to loosen the socialist government reins that have held their economy in check for so long. The phenomenal growth of the Irish economy is another example.

Internet access to the rest of the world and primarily the free world has been a factor in these changes. Even some governments that once controlled virtually every aspect of their people's lives are now recognizing the value of free entrepreneurship, and capitalism. Profit is no longer a dirty word in many of these nations. Tom Friedman details these changes in his recent book, *The World is Flat 2.0.*

Some Predictions

"It is a paradoxical but profoundly true and important principle of life that the most likely way to reach a goal is to be aiming not at that goal itself, but at some more ambitious goal beyond it."

Arnold Toynbee

Studies of energy systems of so many kinds provide evidence that the future will probably see the biggest growth in energy generated in the form of electricity, distributed and used by a wide variety of systems. There will be rapid growth in electric generating capacity. It will be in nuclear primarily, but with geothermal a close second and possibly eventually leading. A decline in coal-fired power plants is probable unless we find a practical technology to gather and sequester carbon dioxide, a difficult and expensive challenge. Wind and direct solar generation still require substantial government subsidies and will remain expensive. They will be only minor players in contribution to the grid. Their use in small, local applications where connection to the electric grid is expensive and to home heating and providing hot water will probably be a substantial benefit and addition to the energy mix. Hydro power will not grow much as environmental concerns will make it increasingly expensive. One novel possibility already producing power, albeit on a small scale, is the conversion of ocean wave action to electrical energy. Thus far, costs and practicality seem reasonable. Downsides appears to be virtually nonexistent.

Vehicles will become more electric and less fuel powered, as battery technology continues to improve, and companies develop better rapid charging systems. There will always be hybrids, mostly electric vehicles with onboard charging capability, because charging capabilities may be unavailable in some places. Of course, there could also be growth in the use of micro turbine generators. Increasing numbers of both remote and emergency power applications are already using these generators for power. The variety of applications will be substantial, as technologies progress and innovators create new ones.

A number of possible alternative fuel and energy systems exist that could replace present coal and petroleum based systems with those that do not use fossil fuels. The benefits of these systems are many, varied, and have far- reaching positive attributes. They run from battery powered vehicles, to biodiesel produced from algae, to a reforming process for animal and vegetable fats, to coal conversion, to liquid fuels. Successful production of these energy systems and fuels would include immeasurable economic and political benefits for the citizens of any state or country that adopts them. This would also garner environmental benefits for the entire world.

There are many innovative new products and technologies that could help us move to a new energy system with a low, or possibly zero net carbon dioxide environmental impact. Many of these are already available and on the market with others are soon to come. This will only be a significant factor if the buying public accepts them and encourages development of even better technologies. Some effective PR would provide a big boost.

Two possible direct replacements for gasoline are butanol (butyl alcohol) and 2,5-dimethylfuran (DMF). Both have been available for a long time, primarily as solvents and paint thinners. Both can be used in current gasoline engines with little or no modifications. Their high cost relative to gasoline is now changing as gasoline prices rise. New manufacturing techniques have already lowered the production costs of these new fuels to competitive levels. Possibly, these fuels could even be made from waste plant materials by active biota. Several new techniques have already shown some success. What we need now is private investment in further research and development of processes that can be scaled up to meet the kind of quantity required for a gasoline replacement.

There are serious side effects from the production of biofuels

Conversion to alternative fuels is already creating serious problems that can only grow worse. Diverting so much corn and soy beans from food production to energy use (as ethanol and biodiesel) is already bringing about major increases in the costs of these grains. Most have hit all time highs on the grain markets and signs of a relaxation of this upward trend are not yet appearing. The rapidly expanding economies of nations like China, India and others are exacerbating this problem. This new bonanza is thrilling farmers everywhere. Food merchants are also pleased as the prices for anything that uses or requires grains—all baked goods, meats, eggs and milk—are already rising quite noticeably in stores.

Another activity that has quite a different but similarly detrimental environmental effect is the growing of palms for palm oil to be used as biodiesel fuel. This brings about the cutting, burning and clearing of much tropical rain forest to grow palms for the highly profitable oil they produce. Destruction of rain forest, with its tremendous capacity to remove carbon dioxide from the atmosphere, could more than counter any gain in the carbon dioxide balance from use of biodiesel made from palm oil.

There are many other ways to produce biodiesel that could become practical were we to pursue them aggressively. One that is well documented is the use of Algae fed nutrients from waste water or other biological waste materials to produce useable oils.

This is an example of just one possible process. Many of these new technologies would bear fruit if some source would provide more research. This would accomplish two needed goals. One would be to produce biodiesel without interfering with food crops. The second would be to make profitable use of waste materials that now cost money for disposal.

Using food crops to make biofuels has already caused disruption to the food supply which is only going to grow worse. Very recently one serious and destructive side effect of the production of biofuels came to light. Rising costs of food products have increasingly priced many foods out of reach of the poorest people in the poorest parts of the world. World Bank food experts estimate that because of this, nearly twice as many were starving worldwide in 2008 than just five years earlier. That makes it even more urgent that we develop energy sources and systems that do not use materials that go into food products and affect their prices. Starvation of millions is a terrible price to pay for noncarbon dioxide producing fuels to satisfy the whims of the global warming crowd.

Because of this we need emphasis on new fuels made from nonfood-chain raw materials along with new battery technologies. Consequently, electric vehicles should garner the most support. An expansion of geothermal power generation to cover the increased energy demand could be our best bet for electric power. These are some of the only readily acceptable and practical options that can lead us away from dependence on fossil fuels. This can be done without a major disruption of our food supply or serious damage to our environment. It is unlikely that any practical development of cost-effective fuel-cell powered vehicles, hydrogen or otherwise, will come to the fore without a major breakthrough in technology. Although such a breakthrough is always possible, there seems to be no hint of any in the foreseeable future.

There are powerful and deeply entrenched economic and political forces all over the world, that actively oppose any system to replace fossil fuels. This is because it would challenge their power and control over energy. Hopefully, our nation will overcome this opposition and lead the world by becoming the first to adopt such a system. If we do not, I'm certain China, India, and several other countries, will jump at the chance to be first with new energy technology and its associated benefits.

What happens when suddenly we have

$200 A BARREL PETROLEUM AND $8 A GALLON GASOLINE ?

If we wait 'til then, it will definitely be too late.

A Scenario of the Future:
a Warning, Hopefully not a Prediction

Dec. 10, 2011—Associated Press: (hypothetical)

Yesterday, crude oil hit a new high on the world market as prices reached and passed the $200 per barrel ceiling. Regular gasoline in the United States is now selling at the pump for $8.19 per gallon. The effect on the world economy and especially the U.S. sector has been devastating. Oil experts see no end to rising oil prices. This is primarily because OPEC is refusing to increase production in spite of rapidly growing demand in China and elsewhere. Iran's Achmadinijad and Venezuela's Chavez have joined forces to dominate OPEC. Their repeated joint impassioned announcements of hatred for America and their plans to destroy our economy seem now to be gaining momentum. Add to this Russia's increasing threats to control and shut down their oil pipeline to Europe. The threat of such an action had an immediate and catastrophic effect on the European oil supply and prices.

The European Union, hard hit by the Russian threat to their oil supply, is processing new travel restrictions for member countries and will wait to see what happens before getting more involved. Officials in these countries have drafted several resolutions and are debating a course of action in the UN. Yet these same officials have made no concrete proposals for action. Perhaps the growing French nuclear energy program is a factor. The expansion of the manufacture of new battery powered vehicles in Germany and England is placing these electric vehicles on the road in growing numbers. Already, EVs are replacing more and more of the fueled vehicles in Europe.

The American delegation has pressed for action on several proposals, but France, Germany, and Russia have opposed every American plan. Strangely, China now seems to be siding with the Americans in urging for action. Perhaps their increasing need for oil has caused them to change their position.

This could happen. Considering current events, this is not an impossible scenario. Wishful thinkers might say so, but supposing a scenario does unfold that causes oil prices to skyrocket (like a series of destructive Gulf Coast hurricanes?). Al-Qaeda and Islamic fundamentalists have everything to gain, and nothing to lose. Their desire to bring chaos and destruction on the West is certainly not diminishing. With Russia now flexing their new economic and military muscles, a new threat has arisen from the ashes of the old USSR. Our dependence on oil is the key to their success and, incredibly, is providing them with vast funds for military weapons, training and recruiting as well as an economic weapon.

Why don't we do something before our economy is totally devastated?

The prices paid for motor fuel are posted on huge signs in every filling station. In spite of this, most Americans understand little of the nuances associated with energy, fuels and the environment or the economic menace these factors pose. I'll wager that long before gasoline reaches $8 per gallon the motoring public will be screaming. Look at the uproar created by $3 per gallon gasoline. They will be screaming hatred at the oil companies and for their politicians to do something immediately! (Even though they have some complicity, oil companies are a convenient scapegoat and recipient of our wrath particularly when public figures constantly condemn them as the cause.) While the best time to take steps is now, before any serious crisis, my guess is that nothing effective will get done until that major crisis inundates us as the current economic downturn might do. That, of course, is human nature. The media could be a big help in this situation, but they seem to lack the courage or ability to do much but condemn.

Of course, there is also the long-range possibility that all the changes from conversion to nonpetroleum fuels and electric powered vehicles would turn out not to be beneficial over the long run. That is a risk we are being forced to take.

What then happens when suddenly we have

$10 A BARREL PETROLEUM AND $.70 A GALLON GASOLINE ?

If this happens, what will be the consequences?

Another Scenario of the Future:
Ten Years into That Future

Dec. 10, 2018 Associated Press: (hypothetical)

Yesterday, crude oil hit a new low on the world market as it dropped to below $7 a barrel. Regular gasoline in the United States is now selling at the pump for $.69 per gallon. The effect on the world economy has been dramatic. Many new alternative fuel manufacturers have gone bankrupt. With consumption now less than half of what it was just five years ago, the resulting glut on world oil supply has dropped the price to unimagined lows. Iran's Achmadinijad and Venezuela's Chavez joined leaders of most of OPEC nations screaming about a Western conspiracy to bankrupt their nations. Announcements of hatred for America and their plans to destroy us seem now to be gaining momentum. Add to this Russia's increasing belligerence because of economic problems and the potential for war is growing. In the European Union, the growth of France's nuclear energy industry, and the batteries and electric cars British and German manufacturers are now delivering, have combined to reduce petroleum use to a small fraction of what it used to be. Offerings from China and even the U.S. are challenging European batteries and EVs. The new American EV Motors Corporation has quickly become the world leader in EV manufacture with the Longest range, 500 kilometers. This was brought about by the development of their efficient new vehicle with in-wheel motors. The revolutionary new U.S. carbon foam lead-acid batteries outperform both Chinese nickle metal hydride batteries and Italian nano lithium ion batteries and are much cheaper. World wide acceptance of these new U.S. products is bringing about an unprecedented economic boom in America after the calamitous crash of ten years earlier.

America is fast becoming the world leader in geothermal power after a breakthrough in drilling techniques developed by American oil drilling companies. This new technique is promising an unprecedented low rate for electricity as more and more geothermal power plants come on line. Previously rates ranged from 7.18¢ per kilowatt hour in West Virginia to 19.8¢ in New York. The new rates range from 2¢ to 6¢ per kilowatt hour less.

Like France, China is relying on nuclear energy because geothermal energy is still too far below the surface there to be utilized economically. Their auto plants are turning out EVs and PHEVs at an astounding rate, most for their growing domestic market. At the same time, they have become the world's largest exporter of EVs. As a result, China has virtually stopped importing oil.

This could easily come true. Changes during the next ten years could easily make this a reality. Such a dramatic reversal of the world's petroleum economy is quite possible. It all depends on which way we decide to move. What we need is an American government that is friendly to creative business, and a public more aware of the reality of how profits create jobs. Such a reversal of form would turn our nation's economy around rapidly. Our wealthy oil companies could change their prospects and make up for the loss of fuel revenue by turning to drilling for heat energy instead of oil. Investing in geothermal power plants instead of refineries could put them in the forefront of the new energy economy.

An article in the October 2009 *Scientific American* magazine described another factor that could lower petroleum costs dramatically. (See the reference on page 11.) All those organizations developing new technologies are not in the fields of alternative energy. Many are in the business of oil exploration and recovery. Their well-financed efforts continue to show major advances and improvement in getting more of the oil that is in the ground, out of the ground economically. These new techniques have already turned once unproductive oil fields into profitable, productive sources of revenue, for those who adopt these new recovery technologies. Do not count the oil business out just yet.

If we do something now before our economy is totally devastated, we will certainly benefit from the changes.

The prices paid for motor fuel are posted on signs in every filling station. Will those change to prices for a quick charge for vehicle batteries? Will we intelligently use existing technologies to move toward this end while going through the probably painful changeover from petroleum fuels to alternative fuels to electric vehicles? Will we weather the storm of new geothermal power plants and the closing of the coal-fired ones? Will we find new jobs for the thousands of people, who will lose theirs as the petroleum industry, and then the alternate fuels industry are displaced and shut down? Will the changes to an electric economy fed by geothermal energy generate enough new jobs? Or will a low price for petroleum keep it the mobile energy source of choice for many years? Only the future will tell. This was just one of possibly thousands of future scenarios, some of which will play out over the next ten years.

A Day in the Life of a PHEV

This scenario shows what could come to pass if one form of the PHEV proposed by the author becomes a reality. It is just one example of many options of a completely new type of vehicle the author proposes be developed. It is the story of an average American, five to ten years in the future. The scenario uses existing technologies and components that are already commercially available. Modification of commonly used designs to adapt to use in small vehicles would be far simpler and much less expensive than what changes a hydrogen fuel cell vehicle system would require.

The Story: Science Fiction, or Soon-to-be Fact?

It's morning, and Sam Wilkes heads out to his garage, his two school-age children in tow. He is taking them to school on his way to his first sales call of the day. Sam opens the garage door, unplugs the charger from the outlet in the garage, gets in, and drives his children to school in virtual silence—no engine noise. The car is operating on electric power from storage batteries. As he drops the kids off, he notices the fuel tank is almost empty, but that is no problem as the battery charge gauge shows 95 percent, almost a full charge. That is enough to move him nearly a hundred miles. On the way to his next stop, he pulls into a filling station and fills the tank with *RN fuel* for $3.10 a gallon. It takes twelve gallons to fill the tank. (See the information about *RN fuel* in the note at the end of the story.)

After completing his first sales call, he heads for his office in Bakersfield to file some reports and pick up several things for his next call. After driving across town, he pulls into a private parking place in front of his office. Noticing his charge gauge shows 70 percent, he plugs in the charger to the outlet in front of his car provided by the company for his use. It will be charging while he is inside. At the same time, solar cells in the roof, rear deck and hood of his car are also charging his batteries from the sunlight as they will continue to do throughout the day. A day parked in the sunlight will provide up to thirty-five miles of travel, free of any fuel or energy costs. Some flow of electricity from the solar cells continues even when it is cloudy although at a reduced rate.

About an hour later, Sam gets into his car and heads south on CA 99 for his next call about 150 miles away. As he heads down the highway, he sets the charger control at 70 percent. The micro turbine will turn on and begin charging when the battery charge drops below 70 percent. It will continue to run until the batteries are fully charged when the micro turbine will shut itself off until the charge once more reaches less than 70 percent, the charger's current minimum setting. The charging system does not start because the batteries are almost fully charged.

Twenty miles down the road, the RN fuel micro turbine powered generator starts up automatically when the battery charge drops below 70 percent. Sam hears a soft whine as the generator gains momentum and finally levels off at its optimum speed. Since he is driving on fairly level ground, the charge gauge slowly creeps up. The micro turbine is generating more electricity than the car is using.

About thirty miles from his destination Sam heads up into *the grapevine*, a steep grade that will take him over a mountain pass at five thousand feet above sea level. Noting his charge gauge is at 95 percent, he knows he has plenty of power to make it over the pass. By the time he tops the pass his charge has dropped to 40 percent. Now he is going downhill. As gravity pulls the car faster downhill, the electric motors stop using power and employ dynamic braking to hold the car at a safe speed—a cruise control in reverse one might say. In dynamic braking, the electric motors become generators, feeding power back into the batteries. The energy that would normally be used up and converted to heat in conventional brakes is instead converted into electric power stored in the batteries. Every time he uses the brakes his batteries receive a charge. At Santa Clarita he turns west on CA 126 for the drive into Ventura where his client waits. As he turns onto CA 126, he notices his charge is back up to 75 percent.

When he stops to pick up his client, Sam turns the micro turbine control off. This will make the car virtually silent as it runs on electric power alone while his client is in the car. Two hours later Sam heads for home. He forgets to turn on and reset the charging system control. Climbing back up the grade, Sam is using a lot of power from the batteries. Several miles below the pass, the charge drops to 10 percent, and the micro turbine generator whirs into life automatically—a safety feature. Sam realizes his error and that he will probably not make the pass since the car is using more power than the generator can supply.

Rather than push his luck and wait until the charge drops to zero, he pulls off into a rest area and parks. The whir of the generator is reassuring as Sam reaches for some papers in his briefcase. He might as well start writing his report while waiting for the charge to be replenished. Solar panels in the roof, deck, and trunk lid, contribute to the recharge even when Sam is driving, just as long as he is in daylight. Fifteen minutes later, Sam notices the charge is up to 40 percent, more than enough to take him over the pass.

Once over the pass, the long, steep descent combined with the generator and the solar panels have filled the charge to 70 percent. By the time Sam is forty miles from Bakersfield, the charge is full, and the generator turns itself off—a fuel-saving economy effort. Sam switches off the system control and drives the rest of the way in silence on battery power only. After parking in his garage, he plugs the charger cord into the electric outlet. The charge is down to 20 percent, but that's fine. Charging from an electric outlet costs just a bit more than a quarter the cost of charging with the RN fuel generator. After dinner, Sam and his wife go out for a show. With the batteries now fully charged, their car has a range of a hundred miles on battery power alone. They will not have to use the generator at all.

Sam's fuel gauge shows he has used just five gallons of RN fuel in driving more than 375 miles. At $3.10 per gallon his fuel cost is $15.50. His cost for electricity from the power plant was about $4.30, making a total of $19.80 or about five cents a mile. During the trip, his micro turbine had emitted little carbon dioxide (by comparison with today's gas vehicles, less than 5 percent) and almost no pollution. If electric power used to produce the methanol component of the RN fuel were supplied by wind, solar, hydro or nuclear power plants there would be virtually no pollution or net carbon dioxide emitted to the atmosphere.

—End of story

About RN (renewable) fuel: The fuel used in the story is a hypothetical, but easily made, infinitely practical, blend of fuels such as: methanol, ethanol, butanol, DMF (2,5-dimethylfuran), and/or other fuels and fuel additives. These are blended in proportions found to have the best cost/benefits ratio considering the manufacturing, transport, and usage. Like M85, E85 and other *flex-fuels*, it could require some minor redesign and/or modifications to many of today's engines. Already, *flex-fuel* vehicles are on the market that would run on some blends of *RN fuel*. Other blends of *RN fuel* could be used in present vehicles without modification. RN fuel is also a nonfossil fuel and as such does not add any net carbon dioxide to the atmosphere. Read more about RN fuel later in the section on fuels in this book.

About solar panels: A significant amount of electrical energy could be generated directly by the sun if Sam had solar panels on his home, or in the roof, rear deck and hood of his car. If he used sunlight to charge the batteries at home or whenever the car was outside, solar panels could provide a significant portion of the car's energy virtually free. It is conceivable that if driven only short distances, to work and for local shopping, most of the energy required would be free. They'd be driving on sunlight. Of course, these panels and their wiring would require a significant investment in both car and home. However, the return on their investment in the form of lowered energy costs could possibly pay for the panels in a year or two of use. That would depend on the amount of sunlight available where the owner lived, drove, and parked his vehicle.

What This Story Could Point to

Is this a possibility? Not only a future possibility but all the technologies used are already available and highly developed. Prototype vehicles such as the one described have already been announced by several car makers and could be on the road in just a year or so. Commercial model availability might take several years, but that would be far sooner than for any hydrogen fuel cell vehicles. Naturally, improvements and evolutionary developments would be involved. Several fuels other than the RN fuel used in the story could be used. These include most all of the fuels described in the section on fuels later in this book. There is also the growing promise of highly efficient and cost-effective batteries making true EVs a practical reality.

With the introduction and scenario setting the stage, we will now delve into the nitty-gritty of energy systems and devices to see where we are and where we could be going.

SECTION III
Energy Systems, Old, New, and Future

I. Fuel Energy Sources

A. Fossil-based Power Source Fuels

Fossil-based fuels are presently our main energy source, as they have been for many years. We mine them from the earth or produce them by refining products we mine from the earth. Whether used to generate electricity, power our vehicles or heat our homes, offices, and factories, they are a contribution of the planet we inhabit. Built up and stored in the earth over millions of years, we are using up these resources rapidly and irreversibly. This has been going on since the first coal-fired boiler started our growing dependence on mined fuels. While still plentiful, petroleum is beginning to show signs of accelerating depletion. As supplies diminish and costs of finding new oil fields soar, prices of petroleum-based fuels will continue their long-range rise, and at an accelerating rate.

These fuels contribute monstrous amounts of carbon dioxide to the atmosphere. Coal, which is mostly carbon, produces carbon dioxide when it burns. Petroleum products make lower amounts of carbon dioxide in producing equal amounts of energy because much of their energy comes from the hydrogen that produces only water vapor when burned. The shorter-chain (and more volatile) hydrocarbons produce the least carbon dioxide with the methane in natural gas, producing the least of all having four hydrogen atoms for each carbon. While coal is in plentiful supply, its use (mainly in power plants) would continue the infusion of carbon dioxide into the atmosphere. We can reduce this release slightly and at tremendous expense by the installation of scrubbers in coal or oil-fired power plants. These scrubbers remove particles, pollutants, and a small amount of carbon dioxide from the exhaust gases as they leave through the stack. The sequestering of carbon dioxide is still an economic problem we have been unable to solve with any practical process.

1. Coal

By far, the oldest and most universally available mined fuel, coal has been used for centuries. The only fuels used before coal—wood, peat, and animal dung—were used for many years in some locales but not in the same vast commercial quantities as coal. Also, they represent renewable fuel resources, and their use does not increase the net carbon dioxide in the atmosphere. Coal represents a big sink of carbon that was originally taken from the atmosphere by green plants. These plants converted atmospheric carbon dioxide into cellulose and other plant materials. The oxygen in our atmosphere was a byproduct. As this vegetation died, it accumulated and was compressed and heated in the ground. The lighter components of this vegetation, were driven off into the atmosphere, leaving mostly carbon and heavy hydrocarbons. When we burn coal in any process, we recombine that carbon with oxygen from the atmosphere, releasing carbon dioxide back into the atmosphere.

The use of coal-fired boilers to generate steam in power plants is a major source of carbon dioxide. We also use coal to make coke for the steel industry. The distillation byproducts of the coke process provides raw material for several major chemical industries. Steelmaking and many other industries add to the production of carbon dioxide released into the atmosphere.

Energy released by the burning of coal in power plants is converted into high-pressure steam. This steam is then used to drive steam turbines and finally the electric generators connected to the turbines. About 60 percent of the available energy in the coal is converted into electricity. The remaining 40 percent is released into the environment as heat. All uses of coal as fuel contributes carbon dioxide to the atmosphere.

Coal is also used in several processes to make liquid fuels like methanol and gasoline. While this is an expensive process, large-scale use would undoubtedly lower costs and our domestic supply of coal is enormous. With the steady increase of petroleum prices, gasoline from coal could become competitive. An article describing the production of methanol from coal is in the appendix. On page 200 see:

Roan, Vernon, Principal Investigator, Daniel Betts, Amy Twining, Khiem Dinh, Paul Wassink and Timothy Simmons. *Final Report—An Investigation of the Feasibility of Coal-based Methanol for Application in Transportation Fuel cell SYSTEMS.* University of Florida, Gainesville, Florida, April 2004

2. Petroleum Fuels

Petroleum, mined from deep in the ground, is the source of most liquid fuels used worldwide. It is easily transported by pipeline, ship, and tanker truck to most places on the globe. It is refined in major manufacturing plants (refineries) into a wide range of special fuels and other chemicals. All uses of petroleum fuels contribute carbon dioxide to the atmosphere. Some uses produce air pollutants, including nitrogen and sulfur oxides along with unburned hydrocarbons.

The petroleum industry is a mature industry with all the inertia of more than a hundred years of doing virtually the same thing. People and systems that are in place in the industry have become deeply entrenched and resistant to change. These entrenchments include strong, long-standing political and economic ties that will be very difficult to move or change. The exploration, mining and extraction part of the industry must be located where petroleum is found. Much of that is under the control of often despotic regimes, including many that seek our complete destruction.

The refining and distribution parts of the industry are more widely spread around the world and are more controlled by broader market and political forces. Environmental concerns have effectively stopped the construction of new refineries in the U.S. For this reason, more and more of our finished petroleum fuels are coming from foreign refineries. This moves more money offshore and subjects more of our finished fuel to controlled sources asking for increasing amounts of our money. There seems to be no sign of a letup in rising prices for all petroleum fuels.

One benefit to at least some parts of our country is that American wells, shut down because they were unprofitable at forty dollars a barrel, are now being reopened. With actual costs now far below the current market price for petroleum, these wells are once more profitable. Some of our oil is now even being sold and exported on the world market. West Texas is again in an oil boom albeit smaller and more limited than that of the 1930s. Wells shut down for years are once again pumping black gold. This is, of course, a temporary respite that will probably run its course in ten or more years. Of course, a serious drop in the world oil market (the 2008 economic debacle) and those wells will again be shut down as unprofitable.

The vast majority of ground transport vehicles now use reciprocating piston engines for power. These are gasoline or diesel engines plus a tiny fraction using other fuels. They consume large quantities of petroleum products and produce tons of pollutants and carbon dioxide. Vehicles based on petroleum fuel and the fueling systems that supply them, are highly evolved with more than a hundred years of technological advancement behind them.

A major political upheaval in Arab oil lands—a crisis with an impact similar to the fictional one described early in this book—is a strong possibility. This would put catastrophic pressure on government and industry to come up with alternatives—fast! The economic pressure of $200 per barrel petroleum and $8 per gallon gasoline would prompt the American public to demand action from their politicians who, together with their media cohorts, would fan hatred for oil companies. While public concern for the environment and global warming is growing, these pressures pale in comparison to true economic forces.

a. Gasoline

Gasoline is the number one vehicle fuel in the world. It is used for autos, trucks, motorcycles, boats, airplanes, tractors, lawnmowers, and countless small powered devices. It is a major source of pollution and carbon dioxide in all these motors. Gasoline contributes sizeable amounts of nitrous oxide, sulfur dioxide and unburned hydrocarbons along with carbon dioxide. It is highly volatile and extremely dangerous—forming explosive mixtures with air.

Chemically, it is a mixture of hydrocarbon molecules with from five to ten carbon atoms. Longer hydrocarbon molecules (fuel oils) are broken or *cracked* by catalytic converters in refineries into the shorter chains of gasoline. These shorter chains make up gasoline that is normally more expensive than fuel oil. Expensive infrastructures are required to make this conversion. Losses during the cracking process contribute to air pollution with hydrocarbons. During the processes of mining, transportation, and refining of petroleum, nearly a fifth of the petroleum is consumed.

Gasoline is a dangerous material to handle, especially in large quantities. Easily ignited by spark or small open flame, it burns rapidly. As the liquid gets hot, the potential for explosion grows. Gasoline fires expand rapidly, and with tremendous heat output. Gasoline fires are notoriously difficult to extinguish. Using water on them is very dangerous and can cause major explosions. It will not extinguish them.

Even the best controlled gasoline engines contribute some pollution into the atmosphere. Small engines used on hand and garden tools are notorious polluters. Fumes from vaporizing gasoline and unburned fuel from open exhausts can be major causes of noxious air, especially in areas like Los Angeles where temperature inversions frequently trap smog and other noxious vapors for days.

Gasoline prices have been going up and down for quite some time. Many factors contribute to these fluctuations. Factors that affect the price of crude on the open market are not the only ones in play. When hurricane Katrina devastated New Orleans and much of the Gulf Coast, the refineries there were severely damaged. At the same time, damages to offshore drilling rigs cut into our domestic supply of crude. This resulted in unleaded regular gasoline rising quickly in price to more than three dollars a gallon at the pump. The combination of a year without damaging hurricanes and a new monster producing well in the Gulf of Mexico combined to bring those same prices down to just a few pennies more than two dollars a gallon. Since then, world demand for petroleum products has soared and with it the price of crude and finished petroleum products. The U.S. price of regular gas rose above $4.00 a gallon in 2008, probably due mostly to manipulation by Wall Street bankers. It dropped quickly to less than $3.00 a gallon in the late summer when oil fell below $70 a barrel on the world market. By mid-November, gasoline was selling at less than $2.00 a gallon, in some places, while crude had dropped to near $50 a barrel. A worldwide recession was bringing about a rapid lowering of fuel prices.

Reduction in available supplies of unleaded gasoline caused by seasonal changeovers, refinery shutdowns, even increases in demand caused by more people traveling can cause gasoline prices to rise even as crude prices are dropping. As a result, many cried that we need to build more refineries. The news was filled with those kinds of comments. Then within a few months, the hubbub died down and will not be heard of until the next disaster. This illustrates how human nature seems to drive us to put off doing things until after a disaster, even in our personal lives. That is especially true when politicians are involved. Stories are legion about dangerous road conditions where railings, stop signs, flashing lights, gates or traffic signals are installed only after a fatal accident occurs at the place, even after many complaints from local citizens. This is merely a different manifestation of the same human trait.

b. Diesel fuel

Though not quite as universal as gasoline, diesel fuel is used in diesel trucks, busses, locomotives, ships, generators and many portable-small devices. Diesel/electric hybrid busses are now growing rapidly in use, especially in California. Diesel engines squeeze nearly a third more energy out of a gallon of fuel than gasoline engines. Though the fuel has more carbon, the increased efficiency lowers the net output of carbon dioxide. The downside is that more nitrous oxide pollution is produced. Diesel fuel is much less volatile and thus safer than gasoline. Diesel engines are generally more complex and more expensive than gasoline engines.

Chemically, it is a mixture of hydrocarbon molecules with from ten to eighteen carbon atoms. It is virtually identical to fuel oil used to heat homes. Several additives are used in diesel to control ignition rates, increase the cetane number, and prevent the formation of noxious pollutants. These are not used in heating fuel. Compared with gasoline, it has about 15 percent more energy content when burned.

Less volatile and thus more difficult to ignite than gasoline, it remains a major fire hazard. Once ignited, it is nearly as difficult to control and extinguish as gasoline. It burns hotter than gasoline. Because of the high efficiency of the diesel engine, it produces less carbon dioxide per amount of energy delivered than gasoline, even though it has a higher ratio of carbon to hydrogen. However, it does produce other serious pollutants including nitrogen oxides (NO_X). Biodiesel is a new type of diesel fuel. Though it can be used in most diesel engines, it is a different material than diesel made from petroleum. Made by several methods from soy beans, corn and even algae grown on waste materials, this fuel is described later in the section on nonpetroleum fuels.

c. Jet Engine Fuel

Similar to, but lighter and more volatile than other fuel oils, jet fuel is used exclusively in jet and turboprop aircraft engines. Burned at high efficiency in high-speed turbines, jet fuel creates less pollution than gasoline or diesel fuel.

Chemically, it is nearly identical to kerosene, a mixture of hydrocarbon molecules with from ten to sixteen carbon atoms. It is slightly more volatile than diesel fuel. In cold weather formulation it uses even shorter carbon-chain hydrocarbons, and is almost as volatile as gasoline.

d. Fuel Oil, for Heating

Several grades of fuel oil are available. The lightest has the same hydrocarbon composition as diesel but is without the additives. Heavier grades have longer-chain hydrocarbons and are less volatile even than diesel or jet fuel. Heating oil is used for home heating and in buildings with oil-burning furnaces. Heat is the only energy product of this use.

e. Heavy Fuel Oils

The heaviest oils from petroleum distillation are used in boilers to make steam to feed steam turbines. This use is mainly in ships as direct power to the propellers and power plants where they power electric generators. The heaviest of these is called Bunker C and is usually black and viscous.

3. Other Mined Fuels

a. Natural Gas

Natural gas is almost pure methane—the smallest hydrocarbon molecule. As such, it has the highest hydrogen and lowest carbon content of any hydrocarbon fuel. When burned, it produces the least carbon dioxide of any hydrocarbon fuel. Used primarily as a heating and cooking fuel in homes, natural gas produces fewer pollutants and less carbon dioxide than fuel oil. Natural gas is used in small quantities to drive turbine generators and to power a few vehicles. It is the most abundant fossil fuel other than coal. Lately, drillers have gone much deeper for methane, suspecting gigantic quantities at depths below two miles. Some drillers have gone as deep as fifteen thousand feet.

b. LP Gas (propane)

A byproduct of both natural gas compression and catalytic cracking in refineries, LP (low pressure) or bottled gas, is primarily propane (three-carbon molecules). It also contains smaller amounts of butane (four-carbon molecules), pentane (five), and sometimes ethane (two). It is stored as a liquid under low pressure of 100 to 250 pounds per square inch. The actual pressure depends on the ambient temperature. Called LP, propane, or *bottled* gas, it is used for heating and cooking in place of natural gas where natural gas is not available. It is also used as fuel in internal combustion engines, primarily in forklift and other applications inside buildings since it does not make carbon monoxide as gasoline engines do. LP gas does, however, generate carbon dioxide and is a source of this gas.

B. Nonfossil-based Renewable Power Source Fuels

These fuels are of two types: those burned as are fossil fuels, and nuclear fuel. There are no other types of fuel. Those that are burned will contribute carbon dioxide to the atmosphere. They will also contribute different kinds of pollutants depending on the application. The nuclear fuels are only used in nuclear reactors and are not burned or chemically altered while providing energy.

Renewable fuels are those that originate from fresh plant material created by photosynthesis. The carbon content was originally taken from the atmosphere and is merely recycled back into it. All energy produced from these renewable sources is stored solar energy used when the plant material was grown. They do not affect the carbon dioxide balance as all carbon dioxide created was just removed from the atmosphere when the plants grew. This is a net-zero effect on carbon in the environment as use of renewable fuels returns carbon dioxide to the atmosphere taken from it during the growth of vegetation. However, any pollution products will also enter the atmosphere and biofuel exhausts are not free from pollution.

1. Ethanol (grain alcohol)

Made mostly from fermented corn and other grains, ethanol is a liquid fuel that is currently being used in a blend of up to 10 percent in gasoline. This reduces gasoline consumption up to 10 percent and reduces pollution slightly. Conversion to total ethanol use is impractical because of its physical properties, especially its high attraction for water and corrosive qualities.

Ethanol is currently getting the most support, primarily from the Corn Belt where new ethanol plants are sprouting up all over to produce ethanol from corn. Financed with private capital by for-profit investors, this growing source of alternative fuel shows much promise for a petroleum-free, carbon-dioxide-neutral energy economy. The problem of concentrating on ethanol as the only fuel raises the specter of the demand eventually exceeding the supply as the available corn crop is limited. Transfer of a substantial amount of corn from the food-chain to fuel production has already resulted in rapidly increasing costs for food—not an acceptable situation. Research of other raw materials for ethanol production is essential if ethanol is to become a major fuel component. Material such as plant waste or even organic waste now going to landfills are prime possibilities.

Ethanol as fuel in E85 can be used in flex-fuel vehicles now appearing on our roads. By combining 85% ethanol and 15% gasoline, E85 solves the problem of ethanol taking up water from the atmosphere. It does require changes in the internal combustion engines to adapt. These changes are not major and *flex-fuel* vehicles are now available that can use both E85 and gasoline. E85 could also be burned in IC engines to generate electricity. Pure 100% ethanol has such a fierce affinity for water it is unusable. It will rapidly draw water out of the atmosphere and form what is called an azeotropic mixture of 95% ethanol and 5% water. This reduces the energy content of the ethanol substantially. It is the highest percentage of alcohol that can be produced by distillation. Using the 15% gasoline substitute for the 5% water increases the power of a gallon of E85 substantially, far more than the 95% mixture with water. Flex-fuel vehicles will run on either E85 or gasoline.

The serious problem caused by grain used for ethanol production is the rapid rise in the cost of corn. Ethanol production has diverted so much corn from food production that food prices have risen even more than fuel. The front-month contract for a bushel of corn (56 pounds) on the Chicago Board of Trade has jumped from $1.86 at the end of 2005 to more than $4 in 2007. That is almost twice the percentage rise in gasoline during the same period. This is great for farmers, but devastating for those who must buy corn to produce foods. It is also driving some ethanol producers out of the market, almost before they get started. These high prices have caused many soy bean fields to be converted to corn with the resulting drop in soy bean production. Because of this, we will have to switch from corn to cellulose (plant fiber) biomass to make ethanol if it is to be a major player in the transportation fuel market. Plant fiber is now discarded as waste. This technology is in need of a boost of development expenditures if it is ever going to be a serious player.

Table Comparing Various Fuels with Gasoline:

Methanol	Ethanol	Butanol	Gasoline
CH_3OH	C_2H_6OH	C_4H_9OH	Many
Energy Content (per gallon)			
63 K Btu	78 K Btu	110 K Btu	115 K Btu
Vapor Pressure at 100 F (Reid V.P.)			
4.6 PSI	2.0 PSI	0.33 PSI	4.5 PSI
Motor octane rating			
91	92	94	96
Air to fuel ratio			
6.6	9	11.1	12-15

2. Methanol (wood alcohol)

Made by several different processes, this single carbon molecule alcohol has fueled many high performance piston engines and has been used in commercial vehicles. The engine must be set up for methanol, that may use much higher compression than gasoline. It is also an excellent fuel for use in micro turbine generators. Methanol has even been used successfully in experimental fuel cells that could conceivably be used in vehicles.

For a period, methanol was the fuel of choice in Brazil where many vehicles were designed to run on that fuel available at the pump. When gasoline dropped significantly in price, many methanol users switched back to gasoline. Now with gasoline prices rising again, methanol is increasing in sales. Methanol is most commonly made from methane, sugar, or coal. Unfortunately, this does contribute additional carbon dioxide to the atmosphere. Methanol can be made from woody plant residues (it used to be called wood alcohol), but can also be made from corn stalks, straw, and other such materials. Made this way, it contributes no net carbon dioxide to the atmosphere as all of the carbon in the fuel originated from atmospheric carbon dioxide. Unfortunately, methanol has an even worse problem than ethanol with water absorption. To combat this, methanol and 15% gasoline are blended to make M85 fuel. This is the same concept used with ethanol. In addition, methanol has even lower energy content per gallon than ethanol. This coupled with the increasing costs of manufacture may interfere with its practicality as a fuel.

Research on a Direct Methanol Fuel Cell has shown some success. It is far too early to know if this will be a viable option, but if it is, it would be infinitely more practical than the hydrogen fuel cell. The outcome of this research is still too sketchy to know if it will ever be economically feasible. It will still have most of the direct costs associated with the hydrogen fuel cell, but without the tremendous infrastructure costs.

3. Butanol (butyl alcohol) a New Player

Produced by processes and from raw materials similar to that required for ethanol and methanol, butanol could be far better than either of these alcohols. Much closer in physical properties and energy content to gasoline than the other alcohols, butanol can be used in virtually any existing gasoline engine. (See the table on the previous page.) Butyl alcohol is rarely mentioned as a fuel in the media as they concentrate on ethanol..

The deterrent of butyl alcohol as a motor fuel has always been the cost. Historically it has been about three times the price of gasoline, but that is now changing. The rising price of gasoline combined with a new process for making butyl alcohol from biomass (corn, plant residue and sugar) has brought the cost down to a competitive level. The big advantage of butyl alcohol over methanol and ethanol is that its physical properties are so close to those of gasoline. This makes it a renewable, nonfossil fuel that can be used in engines designed for gasoline with virtually no changes to the engines.

See this web site: **http://www.butanol.com** for more information.

Environmental Energy Inc. (EEI) holds patent No. 5753474 for a *continuous two-stage, dual-path anaerobic fermentation of butanol and other organic solvents using two different strains of bacteria*. This process makes butanol-hydrogen production far less expensive while improving the yields of butanol to 2.5 gallons and of hydrogen to 0.6 pounds per bushel of corn. The total energy yield will produce as much as 40 percent more energy than an equal yield of 2.5 gallons of ethanol.

Each butanol molecule has four carbon atoms, twice that of ethanol and four times that of methanol. It also contains considerable more latent energy. Made from corn, grass, leaves and agricultural waste—virtually any plant material—butanol is truly a *green* fuel adding no net carbon dioxide to the atmosphere. It also yields no sulfur or nitrogen oxides, or CO. Butanol can be stored, shipped, and dispensed using the same infrastructure as gasoline. It is safer to handle and far less corrosive than either methanol or ethanol. Made by the latest process, its manufacture also produces hydrogen that offers an increased energy yield of 18 percent more than ethanol from the same quantity of corn.

4. Another Fuel Possibility 2,5-dimethylfuran (DMF)

Another possible liquid fuel, DMF, can be made from sugar. While safety and environmental issues of DMF are yet to be investigated thoroughly, its fuel properties and energy content are much closer to gasoline than those of ethanol.

From *www.sciam.com*, the website of *Scientific American* magazine: "Reporting in the June 21 issue of the journal Nature, University of Wisconsin-Madison chemical and biological engineering professor James Dumesic and his research team describe a two-stage process for turning biomass-derived sugar into 2,5dimethylfuran (DMF), a liquid transportation fuel with 40 percent greater energy density than ethanol." See the appendix for more about DMF.

5. Biodiesel

This fuel can be produced from virtually any vegetable or animal oil. Many crops could be used including soybeans, sunflower seed, peanuts, rapeseed, and, again, corn. Even recycled cooking oil is being used. Collectors go to restaurants and pick up their used cooking oil, usually from their deep fryers. After being run through filters and some clarification processes, cooking oil is readily useable in most modern diesel vehicles.

The same applies to biodiesel made from soybeans, palms or algae. Again, demand for the large amounts of raw material for biodiesel could be detrimental to some food costs and raise the price of biodiesel to unacceptable levels. For example, a new biodiesel/ethanol plant was built near my home in Indiana. When it started operating late in 2007, it used a large part of the nearby soybean crop. When it went into full operation later in 2008, it used a large portion of the soybean production of the entire state of Indiana. That one plant has had a major impact on the price of soybeans.

6. Methane

Produced by fermentation of vegetable matter in reactors or even landfills, it is identical to natural gas described earlier. Among easily useable fuels, it is also the lowest emitter of carbon dioxide for any given amount of energy. If made from decomposition of vegetation it emits no net carbon dioxide.

7. Wood

Probably the oldest fuel used by man, it is still the fuel of choice for most of the third world and is used for both cooking and heating often after being converted to charcoal. In some places, so much has been used that it is becoming scarce. Even new growth is cut before any substantial regrowth can replenish the supply. This has contributed to desertification in many parts of Saharan Africa. In America, it is mostly burned inefficiently in fireplaces producing considerable air pollution. It is used also in remote locations or where there is a plentiful supply close by. It is not a significant commercial product in most areas. Recently furnaces have been developed to burn pelletized wood efficiently and with little waste. This is a renewable fuel used strictly for heating in localized areas.

8. Plant Waste

A lot of agricultural plant waste is merely disposed of by any practical means. Some small portion is used as bedding for animals. An even larger portion is burned for heat. For example, sugar cane waste is sometimes used as fuel to boil down the cane syrup to make sugar. Much of it is incorporated into animal food products. By far the most is plowed back into the earth to decompose and add humus to the soil or just thrown away. Some is even being added to wood used to make burnable pellets. It remains a large potential source of underutilized energy.

9. Agni-byproducts

Many agricultural waste materials, that contain sugar and starch, have been used to produce ethanol or methane. In some cases, cellulose is digested by microorganisms and converted into useful materials. These byproducts could be used to manufacture butanol and DMF, but that technology is still in the development stage. Use of waste organic material as a source for biofuel is a needed technology that is in its infancy. The potential to turn these waste materials into biofuel is huge. The biggest problem is the cost of developing the technology and building the required infrastructure.

10. Ancient Fuels—Peat and Animal Dung

Used mostly by individuals for household heating and cooking, these fuels have been used for centuries in many parts of the world. They are not a significant factor in today's energy use.

C. Manufactured Fuels

Combustible fuels created directly by pelletizing wood or crop waste are used primarily for heating. Other fuels are created as a byproduct of manufacturing processes. Much methane produced in land fills from decomposing waste materials is collected and used. These are but a tiny portion of the total fuel usage.

1. Hydrogen

Produced by reforming hydrocarbons or by electrolysis of water, hydrogen was once the fuel of choice in the media spotlight. Recently, the realities of hydrogen as a vehicle fuel have begun taking it out of that spotlight.

A closer look reveals that hydrogen as a fuel has powerful and significant disadvantages unique among all fuels.

It is the only fuel requiring more energy in its creation than could be produced by its combustion or conversion in fuel cells. No matter how produced, hydrogen would consume far more energy than it could produce in any fuel cell. Its manufacture, storage, transport, and distribution would require expensive infrastructures if used in vehicles. The lightest gas in the universe, it requires tremendous compression for practical storage and transport. All vessels and pipes containing hydrogen would be of necessity, quite strong and heavy. This would include the tank on any vehicle. Leaks would pose both a fire and explosion hazard far more difficult to deal with than LP gas or even methane.

2. A New Method to Produce Hydrogen

This revolutionary method could eliminate one big objection to hydrogen—the cost of storage and transport. A totally new and unique method to produce hydrogen has just been developed at Purdue University by Professor Jerry Woodall, a distinguished professor of electrical and computer engineering. (See the section, *Hydrogen from Aluminum and Water* in the appendix at the end of the book.) The information released on May 15, 2007, describes a promising process that could overcome the costly infrastructure required for the transport, storage and distribution of hydrogen gas.

One practical way to use hydrogen and get around the storage and transport problems would be to convert it to methanol in a chemical reaction with carbon dioxide. This process is based on this chemical reaction.

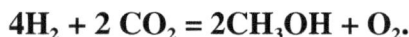

$$4H_2 + 2\ CO_2 = 2CH_3OH + O_2.$$

Four hydrogen molecules reacted with two carbon dioxide molecules, yields two methanol molecules plus one of oxygen. Methanol is infinitely easier and safer to store, transport and use than molecular hydrogen for any purpose. Methanol fuel cells have already been designed and tested.

Note: Almost any liquid or gaseous fuel could be used, with some engine modifications, in properly designed reciprocating engines—even hydrogen. Turbines are even more flexible, as they can be more easily adapted for use with any or even most of these fuels in the same unit. The use of automatic adjustment of the mixture control system could allow turbine engines to adjust for virtually any liquid or gaseous fuel.

3. A New Process Could Produce Viable Liquid Fuels

OXFORD, N.C., Sept. 4, 2007—Catalyzing an entire new industry for North Carolina is the long-term task of the newly established Biofuels Center of North Carolina, which moved to reality today following its first board of directors meeting.

Funded with a $5 million initial appropriation from the 2007 General Assembly, the nonprofit corporation will in coming years implement North Carolina's Strategic Plan for Biofuels Leadership. The Plan was mandated by the General Assembly in 2006 and presented to its Environmental Review Commission in April of this year.

An update: Centia Advanced Biofuels Process Awarded Development Grant from Biofuels Center of North Carolina May 13, 2008—Gilbert, AZ—Centia, an advanced biofuels conversion technology being developed by North Carolina State University, has been awarded a $200k development grant from the Biofuels Center of North Carolina. The Centia process can take any renewable oil input source (e.g., oils derived from agriculture crops, algae, animal fats, waste greases, etc.) and produce transportation fuels that are 1-for-1 replacements for petroleum-based jet fuel, diesel, and gasoline. Fuels produced from Centia could be operated in engines, stored, and distributed identically to fossil fuels today. The process was developed in 2006 by North Carolina State University (NCSU) and has been licensed exclusively by Diversified Energy Corporation.

During this 12-month grant, NCSU will build upon previous test results by demonstrating the end-to-end production of biofuels from a variety of feedstock sources. Starting with one or more North Carolina feedstocks—like crop oils, animal fats, and possibly algal oils—the University will demonstrate all the steps in the Centia process to produce useable quantity batches of renewable diesel, JP-8/Jet A-1 compliant *biojet* fuel, and unleaded *biogasoline*.

The biofuels produced will then be tested in an instrumented single cylinder diesel engine, jet turbine engine, and four cylinder gasoline engine. Dr. William Roberts, professor of mechanical and aeronautical engineering at NCSU and the lead investigator for the grant commented, "The University has been working in earnest to transition this technology from the laboratory to the commercial market. We've had a number of technical breakthroughs within the last year and this funding provides us with an excellent start toward commercialization. With oil at $120 barrel we think technologies like Centia that can produce fungible replacements for fossil fuels from a variety of inputs will be in high demand." Jeff Hassannia, VP of Business Development at Diversified Energy, added, "This award is important to Diversified Energy because it brings recognition from a credible organization that this technology has enormous promise. The Centia process is truly unique and we look forward to its continued technical development and transition to the market to address U.S. energy challenges."

The three steps in the Centia process include:

Hydrolytic Conversion - The feedstock is heated under pressure to separate free fatty acids from glycerol in the triglycerides in the feedstock.

Decarboxylation - The free fatty acids and solvent are heated, pressurized, and passed through a catalyst in a reactor to produce n-alkanes, (saturated hydrocarbons) the building blocks of fuels.

Reforming - Alkanes, straight-chain hydrocarbons of 15-17 carbon atoms (kerosene), are reformed into branched (iso) alkanes and ring structures (benzines). This process is optimized for production of iso-alkanes with 10 to 14 carbon atoms. The alkanes can be reformed differently to create a variety of fuel types. By varying the catalyst, temperature, pressure, and kinetics of this third step, Centia can produce a wide range of biofuels that mimic their petroleum-derived counterparts.

Diversified Energy is supporting the University in providing systems engineering, large-scale plant design, process and economic modeling, and commercialization planning and strategy. The company is also seeking technical and economic partners to support the ongoing development and transition of the Centia technology.

This is just one of many independent research efforts being undertaken by universities and private companies aimed at helping solve this energy crisis.

D. Nuclear Fuels

1. Uranium

Uranium is mined from the earth, refined, and enriched by various processes and then formed into fuel rods or spheres. While no atmospheric pollutants or carbon dioxide are produced, new radioactive elements are created.

Some of these radioactive byproducts are long-lived and quite dangerous. These must be sequestered or permanently removed from the environment. Used fuel rods and spheres can be reprocessed to produce more nuclear fuel. The only practical use of energy from uranium or any radioactive material is in power plants or large ships nearly all of which are military.

2. Deuterium, tritium

Produced by processing seawater, deuterium (hydrogen with a neutron) and tritium (hydrogen with two neutrons), are the preferred fuels for the fusion process being researched by several nations. Successful fusion would produce no pollutants, no carbon dioxide, and no radioactivity. Thus far, practical application of the fusion process has proven elusive if not impossible.

The discovery and development of a practical method of fusion for generating electricity would revolutionize the energy industry. Unfortunately, such a breakthrough seems unlikely soon. Research on the fusion process has been conducted for many years, and the lack of significant progress has been disappointing. Estimates of the time and effort required to succeed in this endeavor have been steadily growing.

3. Exotic radioactive materials

Other radioactive materials can be created in reactors as a byproduct of energy production. Materials, such as plutonium, created in *breeder* reactors can be processed into fuel forms and used in reactors. Unfortunately, plutonium is the material of choice for nuclear bombs and must be handled accordingly.

4. Helium Three

Extremely rare on earth, Helium three is thought to be a possible energy source or fuel in the fusion process. As a result, it is sought for fusion research.

II.   Other Natural Energy Sources

These sources are used as power to convert their energy into electricity by many devices and systems. Except tidal, the ultimate source of all natural energy is the sun's radiant energy. Even tidal energy is partially due to the sun's gravity. Natural power systems add no pollution or carbon dioxide to the atmosphere. Unfortunately, these systems do have downsides, mostly high maintenance costs and the environmental impact of the required infrastructure. Most of these energy systems are used in large installations that produce electricity for distribution via the electric grid.

A. The Sun

The sun is ultimately the source of all energy we use except nuclear, tidal and geothermal. Burning fossil fuels is actually releasing solar energy that originally created these fuels in antiquity. Biofuel energy is solar energy from plant photosynthesis. The book discusses these energy sources earlier. This section deals with energy sources not described in the previous section.

1.   Direct Sunlight

The direct radiant energy of the sun is one of the two most abundant energy sources we have on earth. Geothermal heat is the other. Our direct use of this heat runs from heating water to generating electricity in photovoltaic cells. The sun's energy also drives our climate systems that distribute solar energy in several forms that we also use.

B. Water Power

Water moving from one level to another in rivers and lakes creates power because of gravity. We could tap two other sources with either large or unusual systems.

1.   River Dams

Since prehistoric times, man has damned the flow of rivers and used the energy of the controlled flow of water out of these impoundments. Ancient civilizations used the flow of this water to drive water wheels to provide power to lift water for irrigation and to operate mill stones to grind grains. In many a new settlement, one of the first activities was to dam a stream and build a mill. The energy used comes from the force of gravity pulling the water downstream over ancient water wheels or through modern water turbines. The principle in both cases is the same.

2. Tidal

Tidal forces raise and lower the oceans predictably. This causes large amounts of water to flow in and out of many bodies of water connected to the seas by inlets. Movement of this water can be directed through turbines to generate electricity. Usually we must partially close or dam these inlets to increase the flow rate of the water to produce more electricity.

3. Wave Action

The motion of waves on the ocean holds a large amount of energy, mostly transferred from the wind by friction. Successfully tapping this almost limitless energy is a challenge we are just beginning to study.

C. Geothermal

Trapped in the earth's interior is a virtually unlimited source of heat. Significant amounts lie just a few miles below the surface. In many places, it breaks the surface with varying amounts of volcanic activity. There are two types of interior heat that are useable.

1. Volcanic

This is energy that has broken through the surface or is close enough to the surface that its heat is evident. Many volcanically active areas exist all over the globe. They are usually clustered on or near the edges of tectonic plates. A few are surface manifestations of what scientists call hot spots. These include Hawaii, Iceland, Yellowstone, and Santorini among the best known. The last two are examples of ancient calderas that usually have significant surface evidence of volcanic activity.

2. Deep Heat Energy

Go deep enough under any spot on the earth and things get hotter and hotter as the depth increases. This is the internal heat energy of the earth that is slowly working its way to the surface where it will eventually radiate into outer space. Scientists believe this heat is generated by a combination of radioactive decay and friction. The friction is generated when the moon's gravity tugs on the earth and changes its shape ever so slightly. Many deep mines get intolerably hot at their lowest levels. Some deep drilling for oil must be stopped when the rock becomes so hot the drill head wears out in just a few hours of drilling. The depth at which any level of temperature is reached varies from place to place depending on the thickness and age of the overlying rock. Generally, the heat in the western United States is closer to the surface and thus more easily reached than in the east.

There are many places in North America that exhibit surface evidence of subterranean heat. These hot springs lend the name, *Hot Springs* to many places throughout the nation. Evidently, even in the eastern states there are hundreds if not thousands of locations where the earth's interior heat is quite close to the surface. This is in spite of the fact that with few exceptions, this heat source is quite deep east of the Mississippi. Each of these locations is a potential source for a geothermal power installation. Careful study shows that geothermal energy may be far more widely available than it appears at first glance.

D. Wind Energy

The movement of wind over the earth's surface has tremendous energy as witness the devastation caused by hurricanes and tornados. The size and power of waves, especially along coastlines gives evidence of the transfer of wind energy to the ocean surface. Wind energy is apparent even in light wind. Ask anyone who has ever raised a sail into the wind. Man has learned to harness this energy to power his water craft with sails, to pump his water with wind mills, and lately to generate electricity with wind turbine generators. The winds carry an enormous amount of latent energy. Dutch windmills have harnessed this energy for centuries. The Dutch now use monster wind turbines to generate electricity.

III. Electric Power Plants

Used to supply electric power mostly to the grid, power plants of several types are currently in use. Some of these use fuel and some use natural forces, usually through steam or water turbines. Environmental problems and other dangers and consequences are involved in each type of power generating system. Any energy-generating system we choose will carry a price to pay in money, human discomfort, annoyance and inconvenience.

Increased power requirements: Any new energy system for transportation will probably require a large investment in power-generating capacity, far more than we now possess. These are the present and proposed energy systems of power plants that provide all of that new electricity.

Resistance to changes: Many obstacles to making changes in these systems are evident. Deeply entrenched industries with large payrolls and massive investments are not likely to look with favor on potentially serious competitors. They will do whatever they can to prevent these changes from being successful. It is not because they are evil men, it is a matter of survival—one of our most powerful instincts. To invest in technologies that, if successful, would replace their most profitable components does not make sense for the short run. Unfortunately for these new technologies, the power of the annual report dictates preference for short-term to long-term gains. Entrepreneurs are not nearly so subject to these kinds of pressures and so can benefit from long-range planning. This would be true no matter what the long-term dangers or advantages might be. Only a public outcry like the response to eight dollars per gallon gasoline could change this.

A. Steam Turbine Power Plants

Steam turbine power plants are the most commonly used system for extracting electric power from heat. Water heated to a high temperature and pressure is fed into a turbine connected to a generator, usually on the same shaft. Several types of steam turbines are used. Each can be configured several ways to maximize energy efficiency and produce the most electricity from the available power. The heat used to generate steam can be from combustion of coal, petroleum products, natural gas or even wood. It can also come from a nuclear reactor or even out of the ground where high enough temperatures are available. Once the steam is made, all subsequent processes are identical for converting the heat in the steam into electricity. The power that rotates the turbine is developed by the pressure difference between the inlet steam and the exhaust. Usually a condenser or cooling tower is used to reduce the exhaust pressure and obtain the maximum power.

1. Coal-fired Power Plants

The oldest and most common type of power plant in the world, coal-fired plants still produce more than half the power needs of most nations including the United States. They are also among the most expensive to maintain. In addition, the dangers of mining coal make coal-fired power plants costly to the general health. In these plants, coal is usually ground into a fine powder and blown with air into the combustion chamber of the boiler where it burns at a high temperature. The resulting hot vapor flows around boiler tubes in which the water is heated to a high temperature and pressure. After this, it is piped to the steam turbine where it rotates the turbine, turns the generator and generates electricity.

Coal-fired power plants contain many devices and systems to remove particulate matter, acids and other noxious and corrosive materials from the exhaust gas before it is released into the atmosphere. Before these devices were required, large amounts of both particulate and gaseous pollutants were exhausted creating sometimes deadly air in cities. China experiences serious air pollution because of their rapid industrialization and crude new coal-fired power plants.

Sulfur dioxide from high sulfur coal is still a major problem in our country. Scrubbers to remove it are not installed on all power plants. Those that do have them, find that not all of the sulfur dioxide is removed. This sulfur dioxide in the atmosphere produces acid which results in the acid rain. This corrosive rain has caused many problems in rivers and lakes mostly in the eastern U.S.

One thing that is not removed from the flue gases is carbon dioxide. As this is written, no cost effective process is in use to remove carbon dioxide from exhaust gases. For this reason it is released entirely into the atmosphere. Use of coal as fuel is among the largest contributors to atmospheric carbon dioxide.

2. Natural Gas Fired Power Plants

Similar to coal burners, natural gas fired steam plants directly heat water to drive steam turbines. These plants do not require extensive flue gas treatments as they contain virtually no pollutants or particulate matter. They also release carbon dioxide to the atmosphere though at a lower rate for the power generated than either coal or oil. Natural gas produces about 10 percent of our electricity.

3. Oil-fired Power Plants

Virtually the same as coal-fired power plants, the only difference being the fuel used, and the type of burner and furnace. These must be designed for oil. As with coal, the exhaust gases must be treated to remove pollutants with virtually the same devices and systems. Oil produces only a tiny amount of particulate matter when compared with coal, but also and like with coal, carbon dioxide is released into the atmosphere. About 2 percent of our electricity is produced using oil.

4. Nuclear Power Plants

The vast difference between nuclear power and all combustion processes already described, is in the method of generating the steam that drives the turbines. Once the steam is made, the generating process uses virtually identical turbines and generators. About 22 percent of our electricity in the U.S. comes from nuclear. It currently produces around 11 percent of the world's energy needs.

Uranium mined from various places on the earth is the original fuel for fission nuclear reactors. In a nuclear plant, controlled nuclear chain reactions generate a lot of heat energy. This energy is usually used to heat a transfer medium such as liquid sodium or carbon dioxide circulated through pipes. The reactor uses uranium rods or spheres as fuel, and the heat is generated by nuclear fission. Neutrons smash into the nucleus of the uranium atoms, which split and release energy as heat. The heat from the transfer medium is transferred in a heat exchanger in which water turns into steam at high temperature and pressure. From this point on, the steam is used to generate electricity in precisely the same manner as in all other steam turbine applications.

Nuclear power generation produces no flue gas, no airborne pollutants, and no carbon dioxide. Unfortunately, there is a tiny amount of radioactive waste generated. This can be a problem to transport, store and dispose of. These are not simple problems and have not yet been solved. This is undoubtedly the most serious real problem in the latest generation of nuclear reactors. Perceptions of public safety issues and greatly exaggerated dangers have brought our growth in nuclear power to a halt for at least thirty years. Not everyone will agree with that last statement about the seriousness of the danger. However, France and now China have major nuclear energy growth and are building nuclear power plants at a rapid rate.

A rarely noted fact is that the actual safety record of nuclear power is astounding only because it is never reported. As of 2004, and since 1972, only eight fatalities were recorded per terawatt of nuclear power. During the same period, fatalities directly related to coal-fired power plants per terawatt of power produced numbered 342. Hydroelectric fatalities were 883, and natural gas 85.

Add to the statistics on coal, the indirect deaths from pollution caused by the world's coal-powered stations, and the total is a staggering five million or more each year. So how is it nuclear energy is made out to be the most dangerous by so many people?

5. Geothermal Power Plants

Geothermal energy is the heat held in the interior of the planet. Not far beneath the earth's surface exists a truly inexhaustible supply of heat. This heat manifests itself in volcanically active areas all over the globe. This gigantic sink of heat has barely been tapped for our use. Geothermal may be the long-range answer to cheap and plentiful energy. It is already yielding to our growing efforts to tap into it.

Possibly a sleeping giant, geothermal has the potential to eclipse all other methods of generating electricity. Presently it produces a mere 0.2 percent of U.S. electricity. Used extensively in active volcanic areas like Iceland, New Zealand, and even some in north central California, geothermal may be the best of all worlds. It produces no carbon dioxide, releases no pollutants into the air, and consumes no fuel, so there is almost no cost of transport of fuel. In addition, the cost of a geothermal power plant is about the same as for a coal-fired plant of the same capacity. Once the plant is built, no fuel is used. Maintenance is quite a bit lower for the steam generating section and about the same for the turbine and generator segments.

Potentially, geothermal is the safest, cheapest, least polluting and least carbon dioxide releasing system available for generating electricity. How useful it is depends on several factors: how close to the surface the hot rocks are to start with, how hot they are, and how much water we can pump down to them. In many areas, virtually limitless power is available. As in drilling for oil, expertise in boring deep wells accurately is important because the deeper the drilling goes anywhere, the hotter it gets.

Geothermal energy is not limited to volcanically active areas. An unlimited supply of energy exists (as far as man's use is concerned) at varying distances below the surface, everywhere on the planet. Unfortunately, large areas exist where the heat is just too far down for it to be reached economically using current deep drilling techniques. However, in the U.S., economically viable geothermal power is available over more than 60 percent of the continent, using today's techniques. This opens possibilities in areas not normally considered geothermally active. While this is mostly in the western part of the country, several other viable areas are scattered throughout the rest of the nation.

An article in the October 2007 *Scientific American* reports: "When looking at true costs over a plant's lifetime, geothermal is on a par with or better than a coal plant, the least expensive conventional option." notes Gerald Nix, geothermal technology manager at the National Renewable Energy Laboratory in Golden, Colorado.

Furthermore, an extensive study recently released by the Massachusetts Institute of Technology shows that the heat available underground is surprisingly plentiful nationwide. The report stated, "Geothermal has been dramatically underutilized. Thermally productive rock can be reached by using conventional oil drilling techniques in many areas of North America."

In most geothermal installations, water is pumped down an injection well, filters through the cracks in the rocks in the hot region, and comes back up the recovery well under pressure. It flashes into steam when it reaches the surface. The steam may be used to drive a turbine generator, or passed through a heat exchanger to heat water to warm houses. The steam must be purified before it is used to drive a turbine, or the turbine blades will be damaged and made useless by corrosion. Another way to deal with corrosive water is to pipe it through a heat exchanger to transfer the heat to a noncorrosive medium that then feeds the turbines.

A closed loop or double-concentric piped well would not have that problem. Being a closed system, it would not be subject to the corrosive environment common in volcanic areas. This type of installation does not use ground water, and so avoids the corrosion problem. It can only be used where the hot rocks (above 500 degrees F and below 1200) are stable and not apt to shift and so collapse the wells. There are two types of these. One uses a single well drilled down through hot rocks to a depth that exposes enough well pipe to heated rocks to generate useable steam. The well is lined with steel pipe and then capped at the bottom. Another pipe is then fed down inside the casing with spacers that hold it in the center of the casing. Water is pumped down through the outside pipe where it picks up the heat from the surrounding rock before coming up the center pipe. It is piped to a flash chamber and through a turbine to drive the generator. In highly corrosive rock, the outside pipe is a suitable corrosion resistant material.

The other type requires some precise drilling as two wells are drilled and joined end to end at the bottom. A casing of suitable material is placed inside the entire well system. Water or another suitable liquid is pumped down one well and through the system picking up heat as it passes through the hot rock. The heated liquid comes up the other well, flashes into steam, and drives the turbines. It sounds simple and mechanically it is, but putting it into operation and keeping it running is much more complicated. By the time we have as much development technology on these processes as we do on oil drilling, it could be commonplace.

B. Water or Hydroelectric Power

Virtually all water power is generated by water driven by gravity flowing through a turbine. The pressure differential between the inlet and outlet side of the turbine moves the water. We use several ways to obtain this pressure differential. It is always done by restricting the natural flow of water in some way and forcing it to go through turbines.

1. River Dams

By far the most common type of hydroelectric power plants are those where the power is generated by damming the flow of water in rivers like the Colorado, Columbia, Tennessee and Niagara. Those who have walked on any of these massive structures and seen the giant turbines turning as the water flows through their blades, know the tremendous size of these installations. With initial cost by far the biggest expense, these plants are usually built on a large scale. However, many small, locally known plants, like the Shoshone on the Upper Colorado supply power to local areas. This small plant in Glenwood Springs, Colorado produces up to 15 megawatts of hydroelectric power (enough to power more than 16,000 homes).

Several problems come along with power plants in river dams. These include the ecological changes caused by the dam. Differences in water flow and the level of the impounded water, can make for a variable output. Though infrequent and less costly than with steam turbines, maintenance is still expensive and can take much longer. Because of the large size of some turbine generators, shut downs, even for scheduled maintenance, can be costly as other sources of power must be applied to the grid while the unit is out of service.

Effects of damming on river ecology can be devastating to some fish populations. Many endangered species in the Colorado river are adversely affected by dams that change the flow characteristics of the river, including drastic changes in water temperatures. Serious efforts to reverse these dangerous changes have been and continue to be implemented on the Colorado and other rivers.

The loss of reproductive habitat for salmon, caused by dams in the Pacific Coast rivers of Washington and Oregon, has brought about a drastic reduction in the pacific salmon population. Many ichthyologists believe this salmon population has already collapsed beyond saving while other are asking that drastic measures be taken to improve the situation. These measures, made law for 2008 and 2009, prohibit anyone for fishing for Pacific salmon for at least two years. This has devastated both the commercial fishing industry, and the salmon sport fishing industry of the northwest.

Dams along the Columbia and many other rivers in the area are keeping many salmon from reaching their spawning grounds with the resulting reduction in their reproduction. Combined with what many consider overfishing, this has resulted in a drastic reduction in the number of salmon. In fact, some rivers are now empty of salmon as the ones native to these rivers are now extinct. Ongoing efforts are being made at transplanting newly hatched salmon into those rivers. It is still too early to tell if these efforts are going to succeed.

In some areas, efforts are being made to remove dams and return rivers to natural flows. How effective these efforts will be remains to be seen. These are among the problems negatively affecting hydroelectric power.

2. Tidal

Use of tidal movements of water is similar to that of river dams in that water turbines are placed in the flow path of tidal water flow. Unlike with river dams, the flow stops and then reverses four times each day. Tidal power requires an appropriate estuary or narrow opening where there is a tidal flow that can be dammed. A major drawback of tidal power stations is that they can only generate when the tide is flowing in or out, in other words, only for about ten hours each day. However, tides are totally predictable, so we can plan to have other power stations generating at those times when the tidal station is out of action. Only a limited number of possible sites exist throughout the world. The largest tidal power station in the world (and the only one in Europe) is in the Rance estuary in northern France. It was built in 1966.

Plans have been considered for a Severn Barrage installation from Bean Down in Somerset to Laver Nock Point in Wales. It may have more than 200 large turbines, and provide more than 8,000 megawatts of power (that is more than twelve nuclear power stations' worth). It would take seven years to build, and could provide 7 percent of the energy needs for England and Wales.

Many benefits could result including the protection of a large stretch of coastline against damage from high storm-tides, and the creation of a ready-made road bridge. However, the drastic changes to the currents in the estuary could have major effects on the ecosystem.

3. Wave Action

One of the newest concepts to derive energy from the sea, wave action generators of several types are being studied and developed. Pilot operations are being constructed in several places around the world. These generators convert the energy of moving sea waters into electrical energy by several methods. Buoys tethered to the sea floor move up and down in an elliptical path, turning a generator. Others simply use that elliptical motion and convert it into electricity. In some installations, waves are funneled up the shore through a type of valve that directs the water through turbine generators by the force of gravity as it returns to the sea. Most of these technologies are quite new and unproven, but a tremendous amount of energy is available in the wave motions of the sea. There is an interesting article about an efficient type of wave generator in the July 2009 issue of Smithsonian magazine. Developed by Professor Annette von Jouanne of the university of Oregon, the generator uses magnets attached to a float that slides up and down a relatively fixed stator containing a coil of wire.

The motion of the magnet passing the coil generates electricity just like it does in any generator.

The article can also be viewed on the Internet at **http://www.smithsonianmag.com/science-nature/Catching-a-Wave.html**.

Practical use of this readily available energy store could be another valuable source of electricity. This is now being touted by some as the best possible source of new power with virtually unlimited potential. I do not see it as having any advantage over geothermal at this time.

C. Wind Turbine Power

Wind-driven generators are being built throughout mostly Europe and the western United States, in areas where the wind generally blows steadily. *Wind-farms* on otherwise low-use lands contribute to the electric grid. The fastest growing segment of the energy industry is now wind. Gigantic wind generators are being installed in Denmark, The Netherlands and elsewhere in Europe. Wind is providing a major and expanding part of several nations' electric power. It is now just below nuclear in capacity in Europe. Monster windmills are springing up in many places including out in the ocean. This is to the chagrin of those who live nearby and oppose their spoiling the view. (Shades of Hyannis Port and Teddy Kennedy.) Unfortunately, the best places for wind generators can be far from where the power is needed. This requires long transmission lines and the associated power losses. High maintenance costs and the danger to migrating birds are some of the negatives. The United States lags far behind in the use of wind energy for electricity. The Netherlands, long a stronghold for wind power with centuries of windmills, now has many wind turbines providing electricity.

On a smaller scale, wind turbine generators are providing electric energy for homes in remote areas, for boats and even to power lighted signs in areas without enough sunlight for photo voltaic power. The old windmill-driven pump, once a fixture in every farm is still used to provide water in remote areas.

D. Gas Turbine Power Plants

Natural gas burns fuel directly through turbine drive generators without the intermediate steam process. Used on a smaller scale than coal and in smaller installations, they still contribute to the output of electricity and provide little pollution. Many of these systems are installed to supply power for industrial plants or buildings in areas where natural gas is plentiful or easily mined. They are also used as emergency power in hospitals or other places where power interruptions are frequent or cannot be tolerated.

Very small versions of these called micro turbine generators are now growing in popularity as emergency power for small installations, even some homes.

E. Solar Power

The Sun is the largest energy supplier to the planet. Its energy is generated by the fusion of hydrogen into helium deep in its interior. This energy radiates to the earth supplying us with light and heat. Plants use that energy to make complex molecules called carbohydrates from water and carbon dioxide by the process we call photosynthesis. This process not only produced all the organic compounds we now find in plants and animals, but it also generated the oxygen essential to all animal life and removed all but a tiny portion of the carbon dioxide. We use only a small part of this energy directly in some very different ways.

1. Solar Photo Voltaic Power Cells

Direct solar power using solar cells to generate electricity is the power of choice for small installations in remote places. Solar powered panels supply the energy used in virtually all of our space probes, satellites, and the space station. The cost of transmission lines and connection to the installations in remote places can far outweigh the cost of solar cells. Solar-powered signals and warnings are becoming common along many highways where power is not available nearby. Apparently solar cells have become cheap enough to compete with transmitted power in many areas. With no cost for connection, no meter, and no energy cost, available sunlight is the only requirement.

Though they are still quite pricey, commercially available solar cell panels are now providing electricity for many uses away from power lines. These include RVs, highway warning signs, and water pumps out on the cattle range. Solar panels are even supplying some electric power to homes, mostly in remote areas. In a few homes it even competes with grid power. Solar cells have even been used to power vehicles and are suggested to augment other power sources in some potential vehicles.

2. Radiant Heat Energy

The most common use of radiant solar energy is to heat water for homes. Installed on the roofs of homes and other buildings, pipes or other types of radiant heat collectors absorb radiant energy from the sun to heat water. In some locations, these panels can even provide enough energy to heat buildings and provide hot water for general use.

Radiant energy is sometimes used for commercial drying processes and for desalination of seawater. The advantage is that after the infrastructure is in place, the energy is free. The disadvantage is that in most areas available sunlight varies from day to day and from season to season.

3.   Focused Radiant Heat Energy

In this process, parabolic reflectors concentrate the sun's heat on pipes filed with water. This provides steam that can be used to drive turbines to generate electricity. This type of energy production is also irregular and only practical in areas with sufficient sunlight. At this time it is not a significant contributor of electricity. Normal variations in available sun energy causes dependability challenges as it does with all solar dependent forms of energy production.

IV. The Distribution of Energy

Each form of energy we use requires a different distribution system. Infrastructure required for distribution is designed to carry the specific energy from where it is produced to where it is stored and then to where it is used. The farther an energy usage point is from where it is created, the more the required infrastructure and transport costs. This is true whether it is electrical energy, fuel, or another form of energy.

A. Electricity

The details of configurations of various electrical distribution systems and how they are interconnected are mostly irrelevant to the purpose of this book. This information is readily available on the Internet at many sites and in great detail. Long-range transport via high tension alternating current lines is done at very high voltages to keep losses low. Alternating current is used for most transmission since it is simple to step the voltage up or down using transformers.

1. The Grid, Transmission Network

At the power plant, the transmission substation raises the voltage from between 11,000 volts and 22,000 volts coming from the generator to the extremely high voltage, usually 155,000 to 765,000 volts, needed for long distance transmission. These highly visible power lines run from the power plant transmission substation to power substations. These high voltages are used to reduce transmission line power losses. Power substations have large transformers that reduce the voltage for distribution over the lower voltage lines used for distribution to end users. These local or short distance transmission lines use 7200, 12470, 25000, and 34500 volts depending on their length and how much power they must carry. Typically, the final distribution lines carry 11,000 to 22,000 volts to the transformers that supply power at 220-240 volts to home users. The costs of both the infrastructure and transmission losses vary with the voltage and the distances the power has to travel. Long distances require higher voltages and more costly infrastructure. Both can be reduced as distances become less. Lower voltages can mean substantial transmission losses and the associated costs over longer distances. Power station transformers also incur losses. All these factors are used in deciding where to locate power plants, substations, and transmission lines plus what voltages to use.

Electricity is energy that must be used immediately as it is generated since it cannot be easily stored. Some systems are being used to store electric energy in other forms and so level out variations in the demand. Some water turbines are designed to be used as a pump to fill a reservoir high above the turbine during off-peak demand periods. The water in the reservoir then reverses flow and the turbine becomes a generator for power during peak electrical demand.

2. Batteries

Currently but a tiny portion of energy is stored and transported in batteries. The development of practical battery powered vehicles or EVs could conceivably change that. New technology shows the promise of batteries that can hold and deliver enough electricity to move a vehicle several hundred miles. A few of these vehicles are already on the market. Should they prove viable, electricity stored in rechargeable batteries could amount to a sizeable quantity. With advancing technology, batteries could be one of our most used systems for powering vehicles with almost no pollution or carbon dioxide emissions. That would only be possible if the electricity were generated by nonpolluting power systems. Growth in this area would require expansion of the current electric power generation and distribution grid to adapt to the additional power requirements. That would require considerable investment in new and expansion of old infrastructure—power plants and electric distribution equipment.

B. Liquid Fuels

The infrastructure required for all liquid fuels is virtually identical. The only difference is in the material used that contacts the fuel. For example, petroleum fuels will damage or dissolve some materials used for alcohols and vice versa.

Even some differences between gasoline and diesel fuel could cause problems. Pipe probably moves more liquid from crude oil to gasoline than any other device. Every oil refinery is an enormous collection of miles of steel pipe. Pipe moves the crude oil from the well to storage tanks, from storage tanks to transport (ships, trucks, or long distance pipelines), and from transport to refineries. Pipe also moves finished fuel to distribution terminals, to delivery trucks, to service stations, and even from the service station tanks to the delivery pump and into the vehicle tank. The parts of this system include many items besides pipes including storage tanks, ships, trucks, pumps, valves, flexible hoses and all manner of couplings and connectors to retain the fuel.

Neighborhood filling or service stations are the final and most common part of the distribution network and most of us know them quite well. A few years back filling stations had just two pumps and the associated storage tanks that stored and dispensed regular and premium gasoline. Several additional tanks and dispensing systems have since been added to hold three grades of gasoline plus diesel fuel. Some stations also provide kerosene and even a few new fuels that are beginning to appear. Fuels like E85, a blend of 85% ethanol and 15% gasoline, and biodiesel require separate storage and dispensing equipment. In fact, any new fuel will likely require new and different transport, storage and dispensing facilities. Like diesel a few years back, these new fuels will begin appearing at a few filling stations when the demand calls for them.

1. Fuels Liquid at Normal Temperatures

All these fuels have basic requirements for the strength of the materials used for pipes, tanks and other devices. These materials must be strong enough to hold the internal pressures of pumping and the weight of fuel in the storage tanks. They must also be chemically nonreactive to the particular fuel. This means that the infrastructure, all the pipes, valves, tanks, etc. are inexpensive compared with fuels requiring high pressures. Since everything that holds or carries methanol or ethanol also will contain water, the material used in these pipes and other items must be corrosion resistant. This will add to the cost of any infrastructure.

2. LP Gas (Liquified Petroleum)

These popular fuels require handling equipment that must be strong enough to contain the pressures needed to maintain them as liquids. They also require connectors that seal and hold these same pressures. LP gas is mostly propane and butane. It requires equipment that holds pressures more than 200 psi. This adds considerably to the cost of the infrastructure required for LP gas.

3. LNG (Liquid Natural Gas)

LNG is liquified natural gas or methane with small amounts of other hydrocarbons. To liquify this gas it is cooled to minus 259 degrees Fahrenheit (-161 degrees Celsius). It is liquified and placed in double walled insulated tanks, trucks and ships for transport and storage. Long term storage is accomplished by pumping it underground into places like depleted oil fields. Because of these requirements, the infrastructure required for LNG is much more expensive even than that for LP gas.

C. Hydrogen

The lightest and simplest gas in the universe, hydrogen is among the most difficult and expensive to store and transport. Hydrogen cannot be liquified so any transport or storage has to be under very high pressures. This is done by compressing it into strong cylinders or trailers made of cylinders. All pipes, connections and hoses used for transfer of hydrogen are made for high pressure service. As a result these things are all quite expensive compared with normal items used for liquid transport and storage or even the items used for LP gas or LN gas. Because of this, none of the equipment currently used for fuel of any kind could be used. At present, no infrastructure required for a hydrogen fuel-based economy exists. It would have to be designed, developed, and built in its entirety. That would be expensive.

V. Fuel-powered Systems and Devices

A great variety of fuel powered devices are currently in use. Systems of many types convert fuel into energy and apply that energy directly to wheels or power trains. Some systems convert the fuel energy into other forms—almost exclusively, electricity. The electrical energy produced is then either used directly or fed into the electric power grid.

A. Combustion-Based Systems

Every fuel burned to release energy creates some pollution and carbon dioxide. Fuels like coal have the highest carbon content and create the most carbon dioxide when burned. Fuels like natural gas have the lowest carbon content and create the least carbon dioxide in comparison. Unfortunately, other than hydrogen, all fuels burned contain some carbon and produce carbon dioxide. Renewable fuels return to the atmosphere that carbon dioxide removed during its recent creation. The use of these fuels would result in a no net increase in atmospheric carbon dioxide and stop the current steady increase of that gas.

Working diagrams of each of these engines is pictured and can be viewed at:

http://www.animatedengines.com/locomotive.shtml, on the web.

1. Internal Combustion Engines

These engines burn fuel in a confined space and convert some of that heat into mechanical energy to move vehicles or turn the blades of a mower. Some are used to drive generators to make electricity. They also release many compounds into the atmosphere including unburned fuel now removed by catalytic converters, and most of the following: sulfur dioxide, carbon monoxide, nitrogen oxides, carbon dioxide, water vapor, and some particulate matter.

a. Reciprocating Spark Ignition Engines

By far, the most common energy system used in vehicles throughout the world, is the spark ignition reciprocating or gasoline engine. These engines are inefficient and create many pollution byproducts in most of its applications. The technology of this type of engine is highly advanced, and its manufacture is well established and widespread. As long as practical fuels remain abundant and affordable, this will remain the most popular system for vehicle, marine, small aircraft and portable power sources of all types. This is primarily due to its convenience, low cost, and widely established manufacturing and support systems. The single exception is turbine engines used in large aircraft.

Many types and uses of these engines exist from large vehicles down to the tiny engines on model aircraft. Until turbine engines became the norm for military and then commercial aircraft in the forties and fifties, giant inline and radial engines of as much as 2,000 horsepower, powered these aircraft. Though four-stroke versions of these engines dominate the auto market, the two-stroke versions dominate small outboards, lawn tractors, mowers, hand-held weed whackers, and even model aircraft. Chances are your garage holds several of both types. These engines are major air polluters, spewing unburned hydrocarbons and other noxious products of the combustion process into the atmosphere creating smog and other air quality problems. This is particularly true for those not protected with catalytic converters.

Spark ignition engines can be designed to run on virtually any light liquid or gaseous fuel. They will run on methane, LP gas, LNG, and even hydrogen in addition to gasoline and several alcohols.

(1) A Special Case, LNG or Liquid Natural Gas

T. Boone Pickens is promoting a plan for massive conversion of diesel trucks to LNG, Liquid natural gas. His plan includes wind-farm production of electricity to replace LNG power plants. This change would make large amounts of LNG available to power heavy trucks. California set forth a ballot initiative that would free up $5 billion for deployment of a million LNG vehicles on state roads. In 2006, the ports of Long Beach and Los Angeles adopted a plan to reduce drastically pollution from more than 16,800 Class-8 tractor trailers, the only trucks strong enough to transport the heavy containers in and out of the ports. They chose LNG for many reasons including safety and cost. Even with the expense of replacing diesel engines with LNG engines the ports look to save around $350 million each year. The ports have announced the approval of a new $1.6 billion Clean Truck Superfund. Wal-Mart, which operates one of the largest truck fleets in America, is testing four trucks to measure the possible money saved by the switch. I am certain other truck users are carefully watching the results.

Pickens is positioning himself to profit from this with several investments. By putting $12 billion in Texas wind-farms he will make large quantities of LNG available for trucks. Unfortunately, the economic downturn of 2008 has caused him to put a hold on his wind farm plans. He has a 40% ownership in Clean Energy Fuels Corp. that provides natural gas for vehicle fleets by designing, financing, building, and operating LNG fueling stations. He also has a 12% interest in a small Canadian company that has designed what could be the most advanced, efficient engine on the planet to replace diesels in new and existing trucks. It is powered by LNG.

Pickens is the type of entrpreneurial investor who will probably solve our energy crisis while generating many good jobs and enriching himself and countless others in the process. His efforts will also result in a great deal of tax revenue for several governments. Many other entrepreneurs and planners are investing in research, development and manufacture of advanced energy systems that will help us solve our energy crisis. Unfortunately the combination of promised new corporate taxes and the economic downturn of 2008 has already caused many valuable projects to be scrapped. Pickens and others have downsized or completely abandoned creative new investments in many things including much needed energy and alternative fuel projects. A more encouraging tax policy could reverse this especially if government would just stay out of the way.

For more info go to:

http://www.greenchipstocks.com/articles/clean-energy-fuels/292

or email: **gcreletter@angelnexus.com**

b. Diesel Engines

Diesel engines use compression ignition to ignite the fuel injected inside the cylinder after the air is compressed. The fuel self ignites from the high temperature of the compressed air in the combustion chamber. Invented in 1892 by German engineer Rudolf Diesel, it was patented on February 23, 1893 and named for the inventor. Some confusion exists about the inventor as Herbert Akroyd Stuart and Charles Richard Binney had already obtained a patent (No. 7146) for a compression ignition engine in 1890. However it was Diesel's name that stuck to the engine to this day.

Many good descriptions of the various types of diesel engines are available on the Internet, both two stroke and four stroke types. The big difference from gasoline engines is that they draw only air into the combustion chamber. Fuel is then injected directly into the cylinder right at or near to the highest pressure point when the piston is at its maximum extension into the cylinder. At this point the air has been compressed until it is above the temperature at which the fuel, diesel oil, will ignite as it is sprayed into the combustion chamber. The resulting explosive combustion of the air/fuel mixture raises the pressure and forces the piston down in the same manner as in the gasoline engine. Diesel engines will run on virtually any fuel oil, even vegetable or animal fat of the right consistency. While most diesel oil used today is made from petroleum, biodiesel made from nonpetroleum vegetable oils is growing in production and popularity.

Because of the high pressures developed in the combustion chamber and the high mechanical pressures on crankshaft bearings, diesel engines must be quite a bit stouter than gasoline ignition engines of similar power output. This makes them much heavier than other engines of equal horsepower. This is an advantage for diesel-electric engines used for rail transport as the immense weight in the locomotive provides a surer grip on the rails for starting or pulling up grades.

Diesel engines are the primary power source for most trucks and many ships besides nearly all rail locomotives. A bit more efficient than gasoline engines of the same power, they usually run at a lower rate of revolution. In recent years more small diesels have been made and are appearing in small autos, tractors and other installations. Coupled efficiently with electric generators, they are now used in hybrid vehicles. Diesel/electric hybrid busses are now a common sight in California cities. Running at their most efficient output, diesel engines are ideally suited to run at constant output and maximum efficiency. Presently the most efficient internal combustion engines available, their only drawback is the nitrogen oxides or NO_x they produce that is expelled into the atmosphere. Efforts to reduce these corrosive and noxious components have had some success, but much remains to be done. Development of a diesel engine that would run on lighter, renewable, less polluting fuel would be a beneficial advance.

c. Turbine Engines

Turbine engines have for years been the power of choice for commercial and military aircraft. The large power to weight ratio they have is the best for this application. Continuing development of improving technology has addressed such problems as fuel economy, noise, power control and reliability. Research and development of high-strength metals and ceramics to use in turbine blades has made turbine or jet engines able to rotate faster without rotor failure. Now an extreme rarity, this was once a major cause of dangerous engine failure. This disaster was usually a spectacular disintegration of the rotor when a material failure causes an imbalance—an explosion.

Aircraft turbine engines are of three different types, turbojet, turboprop, and turbofan. All three bring air in through a wide opening at the front where a fan compresses the air, mixes it with fuel and burns it in the combustion chamber. These hot gases then pass through another fan or fans that impart power as rotation to the turbine shaft.

(1) The Turbojet Engine

In the turbojet engine, this power is used solely to compress the incoming air. Power to move the aircraft is provided by the high speed gases as they exit the engine through the opening at the rear. This is similar to rocket propulsion. The higher the speed of the exiting gases, the more power is transferred to the aircraft. Diagrams showing how these engines work can be found on the Internet at:

http://www.animatedengines.com/jets.shtml

(2) The Turboprop Engine

In a turboprop engine the power not only drives the compressor, but it also powers a propeller that moves the plane. Some turboprops use energy remaining in the exhaust gases to add thrust like a jet. Turboprop engines are most efficient at lower altitudes where the air is more dense. Therefore they are used mainly on aircraft that fly at lower altitudes on short trips as opposed to pure jet engines that are more efficient in thinner air at high altitudes and thus for longer trips.

(3) The Turbofan Engine

The turbofan engine is now used in nearly all long-range commercial aircraft. Similar in principal to the turboprop, the propeller is replaced by another fan in front of the compressor blades. This fan provides thrust by moving the air around the jet engine in a shroud that directs the compressed air. This configuration provides power at high altitudes where a propeller would be ineffective and so improves the efficiency over that of the turbojet.

(4) The Micro Turbine Engine

Micro turbine engines have been used for some time to power small generators providing electricity for use away from transmission lines or as emergency power in case of loss of grid power. Many power users like hospitals that cannot withstand a power outage have micro turbine generators for emergency standby power. These engines are slow to change power output, so they have been unsuitable for applications with rapidly fluctuating power demands like vehicles. Since they work best in applications where they are run at a steady speed and power output, they work well in fixed locations and driving generators in hybrid vehicles.

They are just now becoming practical and are being applied to increasing numbers of vehicles both as direct power and as power for generating electricity. Applications for trucks and buses are growing in popularity where turbine generators provide electrical power to both charge batteries and operate electric motors that then power the vehicle. Applications for the efficiency and long life of single shaft, micro turbine generators are rapidly expanding. Micro turbine hybrid systems now used in busses are even being considered for use in other vehicles of many kinds. Autos powered by micro turbine generators and batteries, PHEVs, may soon be available. They await only the development of suitable small sized units.

2. Steam Engines

Steam was the first heat power to be used by man to generate mechanical power. While the old steam locomotive has been replaced by the diesel-electric, the use of steam power is still growing in several kinds of power plants and on ships of many types.

a. Reciprocating Steam Engines

The reciprocating steam engine, once the mainstay of railroads and ships, has become a rarity. Inefficient, dirty, and an obvious air polluter, the old *steam engines* fueled by oil, coal, or wood, have almost vanished. Steamships and a few locomotives are the last remaining users of these engines. The choo-choo of old was powered by steam cylinders moving the huge driving wheels of sometimes monster locomotives. These were the prime overland movers of freight and people in the nineteenth and early twentieth century.

b. Steam Turbines

Used primarily in power plants to generate electricity, steam turbines are also used in some ships. Powered by coal, oil, or gas fired boilers, these turbines drive electric generators or ship propellers. Steam for turbines is also provided by nuclear reactors in some power plants and in nuclear ships, submarines, and aircraft carriers.

3. Stirling Engines

Predating the internal combustion engine by many years, the Stirling engine is a sealed piston unit that runs on heat applied to one end of one piston. Since it is slow to come to power or change power, the Stirling engine is only suitable for use in applications where it can be run at a steady power output. Coupling it with a generator and batteries in a vehicle could be an ideal application. The engine could run at a steady optimum speed charging the batteries while the batteries would provide the varied power requirements for the wheels. A Stirling engine will operate on virtually any source of heat. View a graphic of how this unusual engine works at this web site:

http://www.animatedengines.com/ and select *Stirling* from the menu on the left.

B. Systems Not Powered by Combustion

Systems that fit this description include nuclear reactors, batteries and fuel cells. They do use fuel to provide their energy output, but not in the usual sense of burning fuel.

1. Nuclear Fission Reactors

Using a controlled nuclear chain reaction, nuclear power plants generate heat that ultimately makes steam that drives a turbine just as it does in a coal-fired plant. The latest generation III light water reactors (LWR) are now considered the safest and most popular. For a short time it looked as if pebble bed reactors with their self-regulating property could be the best system. Then it was realized that to be the safest and best, all of the pebbles containing the radioactive material had to be perfect, not a very realistic situation. That and several other realizations have moved this once promising technology out of the limelight.

Used exclusively in electric power plants or in ships, nuclear reactors use the heat of nuclear fission to generate steam for steam turbines that power either electric generators or ships propellers. Radioactive waste products are the biggest problem to deal with after safety. The latest, advanced new light water reactor nuclear power plants are far safer than even coal-fueled or petroleum fueled power plants when considering the entire industry from mining to waste disposal. A lot more information about nuclear fission reactors is available on the Internet. For information, go to:

http://science.howstuffworks.com/nuclear-power1.htm

2. Nuclear Fusion

Fusion is not yet practical, but some exciting new research is being done in Japan. Researchers at Osaka University's Institute of Laser Engineering have been working on a promising new energy technology called *Fastignition Laser Fusion*. This technology uses multiple high-power laser beams to cause a tiny pellet of deuterium-tritium (DT) to implode instantaneously, radically increasing core density and temperature. Then a quadrillion-watt laser fires into the center of the pellet, triggering fusion and releasing tremendous heat to boil water and drive turbine generators.

Fueled usually by deuterium from seawater, laser fusion has the potential to provide plentiful, lower-cost energy from small power plants with virtually no environmental impact. Unfortunately, researchers still have a long way to go. They estimate at least fifty to one hundred years without an unexpected break through in technology.

Forty years ago it was estimated fusion would be generating our electricity in forty to fifty years. In spite of a great deal of research, fusion remains an unsolved problem and though we have learned a lot about it, practical results are probably still forty to fifty years in the future. So don't hold your breath. Like so many of our other problems, a single breakthrough discovery could start fusion on the road to success. It is potentially a safe system with few serious drawbacks. All we have to do is learn how. That may not yield to our efforts for a very long time.

More information on fusion can be found at:

http://science.howstuffworks.com/fusion-reactor.htm

3. Fuel Cells

A developing technology, fuel cells combine a fuel (hydrogen or methanol) with oxygen from the air in a catalytic matrix to generate electricity directly. Fuel cells are efficient and generate little heat or pollutants as byproducts. However, much more research and development will be required to advance these cells into a practical power source for a vehicle. This may never be done economically. One major problem is the amount of money required to develop the infrastructure required to make, transport, store and deliver the hydrogen fuel.

For information on how they work, go to:

http://www.howstuffworks.com/fuel-cell.htm

Regarding the hydrogen economy, Matthew Wald, the *New York Times* science writer reports in *Scientific American*, "Only in an economy like perfume where quantities are so small that price is no factor will the hydrogen fuel-cell be economically feasible."

In another article he wrote the following:

"President Bush has called hydrogen the *freedom fuel*, but hydrogen is not free, in either dollars or environmental damage. The hydrogen fuel cell costs nearly one hundred times as much per unit of power produced as an internal combustion engine. 'To be price competitive, you've got to be at a nickel a watt and we're at $4 a watt,' says Tom R. Dawsey, a research associate at Eastman Chemical Company, which makes polymers for fuel cells.

"Hydrogen is also about five times as expensive, per unit of useable energy, as gasoline. Simple dollars are only one speed bump on the road to the hydrogen economy. Another is that supplying the energy required to make pure hydrogen may itself cause pollution. Even if that energy is from a renewable source, like the sun or wind, it may have more environmentally sound uses than the production of hydrogen. The distribution, storage, and safe handling of hydrogen, the least dense gas in the universe, have many technological and infrastructure difficulties. Any practical proposal for a hydrogen economy must address all these issues."

4. Batteries

A rapidly growing sector of our portable energy economy is powered by rechargeable batteries. After many years of rechargeable lead-acid and nonrechargeable zinc-carbon batteries, several new types have appeared and are growing in popularity and use. The following paragraphs are about EVs (electric vehicles) and batteries. We will take a closer look at the new battery technologies since they are so new and are being driven by the needs of automotive EVs. Pure electric vehicles are the vehicles that use storage batteries to power an electric motor that moves the vehicle. Using older and available technology, most pure electric vehicles are limited to a range of fifty to eighty miles before they must be recharged. This makes them impractical for many people. In spite of this, many are in use, even some built or converted from conventional vehicles by their owners. The development of practical, more powerful storage batteries able to hold much bigger charges may finally be a reality.

A lot of research and development is showing promise along several lines. Many of these are described in the appendix. This may quickly thrust the electric vehicle into the limelight. Should such a change take place, the resulting increase in demand for electricity would require a massive increase in electric power-generating capacity.

a. Lead-acid Batteries

These have been used for more than a hundred years, mostly to power starter motors for internal combustion engines. They have also been used to power small utility vehicles, forklifts, golf carts, power shopping carts and even one quite successful electric car, the Baker electric. This ancient EV commanded 38 percent of the automotive market in 1901, but by 1915, its era was over and production ceased. The lead-acid battery is still by far the most used and most common rechargeable battery.

b. The Firefly Battery

This revolutionary new type of battery, uses lead-acid chemistry, but without the heavy and easily corroded lead plates used in previous batteries. Using a carbon foam support and dispersion matrix, these batteries weigh about half the previous types and have a much higher charge density. Firefly batteries, just now coming to market, show much promise including an all important rapid charging capability. It would be ironic if the same old chemistry we have been using for more than a century would prove to be the best for the new energy era.

c. Zinc-carbon Batteries

These are the first portable batteries. Invented in London in 1834, these cells are not rechargeable and must be discarded. Zinc-carbon cells, for many years the only small battery available, they are still common and the cheapest small batteries. They are available in sizes from D to AAAA. They have also been available in larger sizes for use in lanterns and other large items. They are not practical for use in larger applications like vehicles.

d. Zinc-air Batteries

Cells using zinc-air technology first came on the scene in the early thirties. In 1957 they were used to power Sputnik, the first Russian satellite. In the 1960s they came into use as button batteries for watches and hearing aids. John Cooper at Lawrence Livermore Labs patents the zinc-air refuelable battery in 1993. The battery is charged with an alkaline electrolyte and zinc pellets which are consumed in the process to form zinc oxide and potassium zincate. Refueling takes about 10 minutes and involves draining and replacing the spent electrolyte and adding a new charge of zinc pellets. The short refueling times possible with mechanical charging are particularly attractive for EV applications. The spent electrolyte is recycled.

These cells use the oxygen content of the air to generate electricity. The positive electrode (cathode) is a porous body made of carbon with air access. Atmospheric oxygen from air entering the battery is reduced at this electrode. The active component is not contained in the electrode. It comes directly from the oxygen in the air. The negative electrode (anode) is made of zinc. A water solution of potassium hydroxide is the electrolyte. Although the cell voltage for zinc-air batteries is theoretically 1.65 Volts, they usually are designed to produce 1.4 or 1.3 Volts in order to achieve longer lifetimes.

The zinc-air system, when sealed, has excellent shelf life, with a very low self-discharge rate. They are inexpensive, have a high energy density, but provide relatively low power. For their physical size, zinc-air batteries store more energy per unit of weight than almost any other primary type.

Drawbacks: Zinc-air batteries are negatively affected by extreme temperature and humid conditions. They can be *poisoned* by carbonate, created when carbon dioxide from the air reacts chemically and lowers conductivity. They can have a rather high self discharge rate and the active chemicals tend to dry out. This shortens their useable life so they must be used quickly.

e. Alkaline Batteries

These batteries have all but superseded the old zinc-carbon cells and even zinc chloride in terms of performance. Alkaline cells deliver as much as ten times the capacity of zinc-carbon cells and have superior performance at low temperature. Major brands are now guaranteed leak proof, one of the biggest problems with the early models. They are now the consumers most popular primary cell and are available in many sizes and styles. These batteries are also available as Rechargeable Alkaline (RAM) batteries which are not the same as standard Alkaline batteries. They are not yet practical for use in larger applications like vehicles.

f. NiCad, or Nickel-cadmium Batteries

They were the first rechargeable batteries other than lead acid to appear. NiCad batteries were soon being used in all types of portable, powered devices from flashlights and toys to radios and electric toothbrushes where previously, zinc-carbon batteries were the only ones available. They are plagued by limited power storage, short life and *memory* problems where recharging before they are discharged interferes with and severely reduces their power storage capacity. In their present form, they are not suitable for use in vehicles.

g. Nickel Metal Hydride Batteries (NiMH)

NiMH batteries were the next type developed and were produced early in the 1990s. A bit more expensive, but longer lasting and unaffected by the memory problem of NiCad batteries, they soon became the standard premium rechargeable battery. They are used in most applications today where a rechargeable battery is required. These batteries are even being used as removable small batteries in flashlights, clocks, small gadgets and other applications previously dominated by zinc-carbon or alkaline batteries. Apparently their patent holder will no longer license them for use in EVs. This has been used as the explanation for why GM and Toyota scrapped their successful electric vehicle programs in the face of rapidly rising demand.

h. The Lithium-ion Battery

They are low-maintenance batteries, an advantage that most other chemistries cannot claim. Unlike NiCad batteries, they have no memory problem, and no scheduled cycling is required to prolong the battery's life. In addition, the self-discharge rate is less than half that of nickel-cadmium batteries. This makes lithium-ion well suited for modern battery applications. Lithium-ion cells cause little harm when disposed. Used mostly in laptop computers, tools and similar installations they are just becoming available in standard cell sizes. On rare occasions, they can have overheating problems leading to difficulties including even explosions in larger sizes. Application of nanotechnologies to the lithium-ion battery has resulted in a much more stable and longer lived battery. It also makes possible batteries that will accept a fast charge—about ten minutes for a full charge. Several companies are already bringing these batteries to market for EVs. The details are in the appendix. As volume goes up, prices should drop considerably.

i. Nickel-ferrous Battery

Another new technology that shows much promise is the environmentally friendly nickel-ferrous (NiFe) battery. It will compete with lead acid, nickel metal hydride, lithium, and zinc-air, on a cost-per-watt-hour, performance, and weight basis. With development both here and abroad underway and production models scheduled to be released as this is written, NiFe technology could prove to be the most cost effective. These batteries are already being built for use in hybrid and electric vehicles. In fact, China's BYD, a major producer of laptop batteries, has just debuted the first nickel-ferrous battery-powered dual-mode hybrid vehicle. The battery is recyclable, has no heavy metals and is capable of a fast charge. In just 10 minutes, the car can be charged to produce a 68-mile ride.

C. Hybrid and Combination Systems

1. Diesel/electric Combinations

These are used in nearly all rail engines. They are also used in ships including some submarines. Some giant earthmoving vehicles also use this for power. Many busses, particularly in California, are now diesel/electric hybrids. They are currently being developed on a small scale for use in autos.

2. Dual-electric buses

Presently uncommon, such vehicles can run on electricity from trolley wires using pickup rods or on diesel-electric power when trolley wires are not available.

3. Gasoline/electric Combinations

These have been available for some time in several small autos, the *hybrids*, made by Toyota, Ford, and Honda. Others, including American manufacturers, have recently begun delivering their own versions of not only small sedan hybrids, but SUV hybrids as well. A few PHEVs based on this power source are coming to market in a few years. The Chevrolet Volt is one of these types and is scheduled for introduction as early as 2010.

4. Gas Micro Turbine/electric

Used for both main and stand-by power for increasing numbers of buildings and remote locations, these compact units frequently have costs lower than that of power from the electric grid. This includes small, even portable power plants. Micro turbine-powered, hybrid vehicles are showing up all over the world in increasing numbers, as they grow in popularity with bus and truck owners. China in particular is buying American built micro turbine generators for their busses.

5. Nuclear/steam Turbine Power Plants

These installations generate electricity in power plants and some ships. They are a major source of electricity for the electric grid. Unfortunately, anti-nuclear activists have stopped all nuclear power plant construction in the U.S. for the last thirty years. France has been building nuclear power plants for several years. They now sell power to other members of the European Union who will not build nuclear power plants.

6. The Hydrogen Fuel cell Vehicle

This has been hyped as *the answer* to the energy crisis. In this vehicle, a fuel cell converts energy from the oxidation of hydrogen into electricity, which powers the vehicle. Only a few prototypes have been created. This type of vehicle is the basis for the so-called hydrogen economy. For many reasons, the hydrogen fuel cell is unlikely to be a practical system for use in vehicles. The chief detraction is the immense cost of the infrastructure required to manufacture and distribute hydrogen.

VI. New Hybrid and Other Vehicles Now Available, or Soon to Be on the Market

Many new and different types of vehicles now being delivered or soon will be. These include hybrid vehicles from many popular auto manufacturers, some radical new EVs, and even some PHEVs (Plugin Hybrid Electric Vehicles). While one of these radical new vehicles is coming from GM, most are from small, private, start up companies, trying to break into the market. After hybrids, whose ranks are increasing geometrically, pure electric vehicles or EVs are the most common. These include both new vehicles and several types of conversions of existing hybrid vehicles. Without exception, these small manufacturers have long waiting lists of people who have ordered vehicles, many even paying in advance. Most are quite similar in look and function to existing vehicles while a few have a radical appearance.

In this section, we will not include the specialty vehicles that are impractical for highway travel. Excluded are many electric vehicles rented to tourists in vacation areas, golf carts and other short trip, usually open air electric powered vehicles. Also excluded are vehicles in the idea or drawing board phase. These vehicles usually have little chance of reaching production for sale. All vehicles described are either already on sale or scheduled to go on sale in the foreseeable future. Vehicles from any established manufacturer will qualify if it is announced they will be produced for sale by a certain date.

A. Hybrid Vehicles

Now the most popular new type of vehicle, hybrids are available from many manufacturers in a variety of styles and models. Toyota and Honda were first to offer viable hybrids for sale. The Toyota Prius soon became the hybrid of choice and is now by far the most common hybrid on the road.

Following is a list of many 2009 hybrid cars and SUVs along with their reported fuel mileage, listed highest mileage first.

Position, cars	City/hwy	Position, Trucks	City/hwy
1. Toyota Prius Hybrid — 51/48		1. Mazda B-Series Truck — 21/26 (tie)	
2. Mazda Tribute Hybrid — 34/31 (tie)		Ford Ranger — 21/26 (tie)	
Ford Escape Hybrid — 34/31 (tie)		2. Toyota Tacoma — 20/26	
Mercury Mariner Hybrid — 34/31(tie)		3. GMC Sierra Hybrid — 21/22 (tie)	
3. Saturn Vue Hybrid — 25/32		Chevy Silverado Hybrid— 21/22 (tie)	
4. Toyota Highlander Hybrid — 27/25		4. Nissan Frontier — 19/23 (tie)	
5. Jeep Compass — 23/28 (tie)		Suzuki Equator — 19/23 (tie)	
Jeep Patriot — 23/28 (tie)		5. Chevrolet Colorado — 18/24 (tie)	
6. Mazda Tribute — 22/28 (tie)		GMC Canyon — 18/24 (tie)	
Ford Escape — 22/28 (tie)		6. Dodge Dakota — 16/20	
Toyota RAV4 — 22/28 (tie)			

These hybrids use similar engine/generator systems with a battery that will only power the car for a short distance. In effect, their generator/motor system acts nearly the same as an automatic transmission in a conventional vehicle by providing the extra power needed for acceleration or climbing hills. The gasoline engine in these vehicles is not much smaller than conventional engines in the same size vehicle. The biggest factor in their fuel economy is that when the driver, *gives it the gas*, much of the power comes from the battery and not from a wide-open throttle. Change any hybrid into an EV by adding a much larger battery, and using the engine only to power the generator to charge the battery. Separating the engine from the drive train completely completes the change. All power to the wheels through the electric motor or motors then comes directly from the battery. The next step up from this EV with onboard charging is simple and easy. Add plugin charging capability and it becomes a PHEV or plugin hybrid electric vehicle. That seems to this writer as the natural next step in automotive evolution.

B. PHEVs or Plugin Hybrid Electric Vehicles

As of this writing, no PHEV has yet come to market, but both Toyota and GM are racing to be the first to sell one. An excerpt from an article by John Addison on April 17, 2008 follows. The complete article can be viewed on the Internet at:

http://www.cleantechblog.com - search on *GM EV*

Electricity is the most promising alternative fuel for GM and most auto makers. Electric motors are far more efficient than gasoline engines. Electric motors are used in hybrids, plugin hybrids, battery electric vehicles, and hydrogen fuel cell electric vehicles. In late 2010, General Motors will start selling the Chevrolet Volt, a plugin hybrid. This will give many drivers 100 miles per gallon of gasoline, because it will run primarily on electricity. In three years, consumers may have multiple plugin choices including Toyota's planned offering. The Volt is an implementation of E-Flex. GM's E-Flex is an electric drive system centered on advanced batteries delivering power to an electric motor. Additional electricity can be delivered by a small engine coupled to a generator, or by a hydrogen fuel cell. In the future, GM could elect to implement E-Flex in a pure battery-electric vehicle. Thanks to the early success of hybrids, more than two million vehicles now use electric motors and advanced batteries. Electric drive systems will continue their strong growth as they are implemented in battery electric vehicles, hybrids, plugin hybrids and hydrogen fuel cell vehicles. The plugin hybrids' big competition will be battery electric vehicles (EV). London's congestion tax is cascading into a growing number of cities that will require zero-emission vehicles. Announced EV offerings are coming by 2010 from Nissan, Renault, Mitsubishi, Subaru, and emerging players such as Smart, Think, Tesla, Miles, and a host of Asian companies, that will display at the upcoming China Auto Show. With the average U.S. household having two vehicles, these EVs would be perfect for the 80% of U.S. driving that requires far less than 100 miles per day. My guess is that the PHEV will be the vehicle of the future since it provides the economy of a battery powered electric vehicle with the flexibility of a fueled vehicle. While pure electric vehicles are now promising future ranges comparable with current vehicles, the convenience and flexibility of an onboard power source, using easily transported liquid fuel, assures continued sales of PHEVs. One downside will be the requirements for a greatly expanded electric power generating capacity. The questions remain: can we build the power plants required in time, and what kind of generating systems will we decide to use in these new power plants?

VII. Energy Systems and Devices Not Powered by Fuel

The only nonfuel vehicles that have ever been available are pure electric vehicles or EVs. These have been around almost as long as gasoline powered vehicles and between 1910 and 1915 they were quite popular especially with women. They liked the fact that they did not have to be cranked, they were quiet, and did not smell so badly. The Baker Electric made in Cleveland Ohio from 1901 to 1915 was by far the most popular though expensive compared with gas buggies. At one time they held almost 40 percent of the auto market. After 1915 sales plummeted and the Baker company went out of business a few years later. Baker electrics were still common in cities into the 1940's. When in high school, one of my jobs was to drive a neighbor lady around for shopping and to visit friends in her car. It was a Baker electric and it was fascinating. This ancient (30 years old at the time) vehicle was impressive. For more information go to:

http://www.lkwdpl.org/lore/lore197.htm

A. EVs, or Pure Electric Vehicles

EVs have been around for a long time although no popular EVs were produced for most of the last hundred years after the Baker electric was number one. An excerpt with information about new electric cars late in the twentieth century follows. It was taken from an article by Santa Monica Mirror staff writer Lynne Bronstein. The entire article about Zan Dubin Scott and her electric RAV4 can be viewed on the Internet at:

http://www.smmirror.com/MainPages/DisplayArticleDetails.asp?eid=7630

Santa Monica resident Zan Dubin Scott believes in electric vehicles. She and her husband Paul bought a Toyota RAV4 back in 2002 and have been driving it for six years. "We haven't been to the gas station once in that time!" says Scott. "And we charge our car with electricity generated by the solar panels on our roof. We are driving on sunshine." But the auto industry hasn't produced electric cars lately. As chronicled in Chris Paine's film, *Who Killed the Electric Car*, [2] the auto companies had originally manufactured about 5,000 electric vehicles as a response to a California state mandate to produce a certain number of zero emission vehicles. The companies, however, feared a loss of profits and began to dismantle the electric vehicle program—literally crushing the electric cars. Toyota, in fact, stopped its electric vehicle program a week after the Scotts had purchased their RAV4.

Protests (shown in Paine's film) were organized, and the Scotts were among them, helping to form **don'tcrush.com**. Now they have co-founded a new group, Plug In America, which is seeking to bring back production of electric vehicles. Paine is appropriately chronicling the events for a sequel to his first film, to be called Who Saved the Electric Car?

EXCITING LATE BREAKING NEWS!

Toyota announces it will start producing an all electric RAV4 in 2012 in cooperation with Tesla Motors of California.

An excerpt from the WIRED article, ***Toyota, Tesla Resurrect the Electric RAV4*** By Chuck Squatriglia Email Author, July 16, 2010. To read the entire article go to:

http://www.wired.com/autopia/2010/07/toyota-tesla-RAV4-ev/

Toyota is essentially updating an electric vehicle it built in limited numbers between 1997 and 2003. The vehicle featured a 27.4-kilowatt-hour nickel-metal hydride battery that recharges in 5 hours at 240 volts. Toyota leased RAV4 EVs to utilities, businesses and cities through 2002 and sold 328 of them to private citizens in 2003. Unlike other automakers (cough GM cough Honda), Toyota didn't crush the cars at the end of its EV program, and most are still on the road.

"I'm still getting 120 miles of range," said Paul Scott, a founder of the advocacy group Plug-In America. He bought his in September 2003. "It still runs exactly the same as the day I bought it. There's no degradation of the battery, no variation in the performance. It just runs."

What's more, the original RAV4 was, and remains, an impressive vehicle. Toyota could slap the original's electric drivetrain in the current RAV4 and have a competitive EV. That's a testament to the original's technology and Toyota's EV expertise.

Another eye opening article about EVs by Sarah Wendorf in the Sudbury Star follows. It is no longer available at the original address, so we cannot provide a link. It might be found in a search at:

http://www.thesudburystar.com/

Does anyone remember the electric car? It was among some of the first vehicles made in the 20th century. (The Baker electric described earlier) General Motors made one called the EV1, introduced in 1990 and put into production in 1997. The EV1, with the battery technology from the late 1990s and early 2000s, could reach up to 240 kilometers per charge and reached an electronically limited top speed of 130 km/h. With today's lithium battery technology, these distances can surpass 300 kilometers per charge, and estimates found that 90 percent of people who drive would be within this range for one day's charge.

As well, the car was not some kind of science-fiction experiment or strange-looking piece of machinery—it happened to look just like today's subcompact cars. The EV1 had a simple recharging system where one would just plug the car into any electrical socket. About four hours later, the car would be fully charged. Around 1,000 EV1 cars were produced for consumer use in the United States from 1997 to 1999.

Shockingly, in 2003, General Motors pulled the plug on the EV1 program. Many believe this was caused by pressures from oil companies for obvious reasons, the U.S. government (loss of fuel tax revenue), auto manufacturers (low maintenance and parts replacement costs meant lower income), consumer misinformation, and other factors. GM argued there was a lack of customer demand for the electric car and that people were unwilling to pay higher prices for lithium batteries. That was pure baloney as researchers found that customer demand was high and that with mass-production, the cost of the batteries could be reduced. The urban legend about the accidental sale of the hundred-mile-per-gallon carburetor to someone's cousin or friend has been circulating at least since WWII. This electric vehicle subterfuge, however, is very real.

It was not just GM that created electric cars. Honda, Toyota, Ford, Nissan and others also created their own versions and, strangely, all of those programs stopped in the early 2000s as well. Did anyone else know about this? Some serious and time-consuming research revealed that almost every major car company had an electric vehicle in production not that long ago. I would like to know how many people would be willing to purchase an electric vehicle in Sudbury knowing that you can have more than 300 kilometers in one charge and the ability to reach the same top speeds as every gasoline-powered vehicle out there.

Several startup manufacturers have recently reported they will soon begin deliveries on new electric vehicles (EVs) that are quite different from the stodgy old Baker. With ranges up to 200 miles and adequate speeds for highway or freeway, these range from the sporty Tesla at around $100,000 to the utilitarian Phoenix pickup for half that amount. Ebon Musk, cofounder of PayPal, made a fortune when the company was sold to ebay in 2002. Tesla Motors is one of his new start-up companies. The San Carlos, California company manufactures the Tesla sports car with deliveries scheduled for the summer of 2008. They already have a considerable backlog of orders. Musk promises their second product will be a sedan at half the price of the sports car and should be available for purchase in 2010. Other new EVs are already available. AC Propulsion offers a conversion of Scion's xBox into what they call the eBox. ACP removes the existing engine system and replaces it with a battery and electric motors and controls for around $70,000. This vehicle is now on the market and on the road. It uses new technology lithium ion batteries and provides a fast charge option in addition to the standard overnight charge.

Myers Motors NMG, (No More Gas) personal vehicle made in Talmadge Ohio is a small specialty vehicle that holds but a single person. With a top speed of seventy-five miles per hour and a range of about thirty miles, this vehicle provides safe, sure transportation for one at about two cents per mile. For those who drive to work or go shopping locally alone it could be the answer to low-cost, noncarbon dioxide emitting transportation. Priced less than $30,000 it is the least expensive EV available. Myers Motors executives are very excited about the latest developments in 2010. Here's what they have to say:

No more Gas. Doesn't use Oil. We have been looking to provide and many of you have been looking to buy a vehicle that doesn't require the use of oil for many of your daily travel requirements. Our goal at Myers Motors is to provide this in a package that is affordable to buy, practical to use, fun to drive, and fun to be seen driving. For most people, we hit at least 3 out of 4 of these targets with our current NmG.

Our goal was to get enough pre-sale interest to justify a profitable sales price of $24,995 for our two-passenger Duo. I am happy to announce that between pre-order interest and some engineering improvements, we are within $1000 of that figure. We believe we will be able to meet our target pricing by the time our Duo hits the street.

Due to engineering changes that, we believe, will make the Duo even better, production will likely be delayed until 1st quarter of next year. I hope you think that the unique styling, no-rust chassis body system, two passenger seating that covers the occupancy needs for almost 9 out of 10 trips made in the US — and which allows the vehicle to be light enough (while being strong enough) to be charged at night from our nation's existing infrastructure at standard 110 volt outlets (and not requiring special chargers) is worth waiting for. With its limitation it would probably only work as a second car for those close enough to work and shopping to get there and back on a single charge. Use of new high-tech batteries for greater range and quicker charging coupled with expansion to hold two adults would make this interesting vehicle much more attractive to most people.

For more information go to: **http://www.myersmotors.com**

Another new kid on the auto block is the Aptera Typ-1e. This is an electric car that looks like a space ship. They expect to market a hybrid version in 2009-10. The electric Aptera has a range of 120-miles on a full 8-hour charge. Its efficiency is based on its body shape. The body was designed to slip through the air with the least resistance. The designers reached a drag coefficient of just 0.15, the slickest shape ever in a car. At the low price of about $30K it looks to be very attractive, if it lives up to the manufacturers claims. The car is designed for two passengers, but a four-person version is in the works.

http://www.autobloggreen.com/category/aptera/

Popular Mechanics has named the Aptera Typ-1e the biggest automotive breakthrough of 2008. Read about this at:

http://www.popularmechanics.com/science/research/4286850.html?page=2&series=60

Quite a few EVs are currently in the works all over the world including in China, Japan, England, Germany, and Russia besides in the U.S. To check some of these out, go to:

http://www.autobloggreen.com/ and search on PHEV or plug-in EV

Chevrolet plans to enter the market with their Volt concept car introduced at the Detroit auto show in 2007 and presently expected to have a production version ready in 2010. Not a true EV, the Volt carries a small gasoline engine driving a generator that charges the batteries while on the road. Plugin charging capability is also included making the Volt a true PHEV or Plugin Hybrid Electric Vehicle. It remains to be seen if Volt lives up to its press clippings.

The last 2007 issue of Newsweek contained a small article about GM's Bob Lutz and the Chevy Volt titled, *The Man Who Revived the Electric Car*. In the article Lutz admits that he realized Tesla Motors, a tiny Silicon Valley start-up, announced in 2006 it was to begin producing a speedy sports car powered by lithium-ion batteries. "That tore it up for me," says Lutz. "If some Silicon Valley start-up, can solve this equation, no one is going to tell me anymore that it's impossible." Lutz must have missed several other start-ups that are already producing battery-powered EVs. Most of these are now on the market with long waiting lists of customers. Some of the new vehicles are described in the Appendix and include the Tesla roadster, Phoenix Motors SUT, Myers Motors NMG, Aptera Typ-1e, and AC Propulsion eBox. All are true electric vehicles and all have electric ranges of 100 miles or more (except the NMG) and top speeds in excess of highway requirements. Unfortunately these companies do not have the large advertising budget of a Toyota or GM so TV, Newsweek and other media publications completely ignore them. More information on these EVs and other new vehicles is available in sections VI, VII, and in the Appendix.

B. Battery-powered Small Vehicles and Tools

Many small, local-travel, battery-powered vehicles are in use. Golf carts and, fork lifts have been in use for quite some time and are still probably the most numerous. A search of the Internet for neighborhood electric vehicles (NEVs), will turn up dozens of makes, types and models. These small, battery-powered, three or four wheeled cars have been for rent in many tourist locations for years. Battery-powered shopping carts, wheel chairs, even scooters are now common everywhere. With their short ranges and occasional uses, these have conventional lead/acid batteries and are plugged into chargers when not being used. The new battery technologies are sure to be used in these applications soon.

Increasingly we see small cordless hand tools powered by NiCad and now Lithium ion batteries. From small drills and screwdrivers using AA batteries to bigger yard tools and chain saws, these new tools use both standard and proprietary battery packs for hours of use. Cordless tools are rapidly replacing tools with cords and even some gasoline-powered yard tools.

C. There are a Few Other Vehicles Not Powered by Fuel

1. All Electric Trolleys

These and *trackless trolley buses* use several types of electrical contacts to take power from the electric grid. The usual connection to the power grid is with spring-loaded pickup poles and dual overhead wires.

2. Inertia-powered Buses

These buses use a heavy flywheel to store energy for short runs. The flywheel is powered up by an electric motor that becomes a generator to provide electricity to power the bus drive-motor.

3. Air-powered Vehicles

These vehicles use a high-pressure storage tank for air to power an *air motor* with no requirement for batteries, fuel or combustion. The *motor* in these vehicles operates in the same manner as those used in pneumatic tools. Theoretically these vehicles could be filled with air by compressors at conventional service stations. Power for such vehicles ultimately comes from the electric grid. Carbon dioxide gas is sometimes used instead of air in these types of vehicles.

D. Where does all this energy originate?

With virtually no exceptions, the power to charge all these batteries comes from the electric grid. This means that should we convert to battery power, our electric power plants would have to roughly double their present output. As far as fuel use, pollution, and carbon dioxide emissions are concerned, this conversion would move the emissions from our vehicles to the power plants. Unless we change to emission-free power from coal-fired that would be the case.

VIII. Fuel Pricing and Other Factors

In an early manuscript written for this book in 2003, the scare page read, *What happens when suddenly we have $100 a barrel petroleum and $4 a gallon gasoline?* In 2005 that was changed to $200 and $8. Today with petroleum fluctuating between $50 and $150 a barrel and diesel fuel from $2.75 to $4.50 a gallon, even reaching those astronomical values may be but a matter of time. By calculation based on the last doubling in price, and unless something changes, petroleum will probably be $200 a barrel, and gasoline will be $8 a gallon by 2012. This will be driven by speculation and deliberate price manipulation. That date depends on when the world's down economy begins to recover.

Gasoline prices have been going up for quite some time with a few periods of minor reductions. Many factors contribute to these fluctuations including speculation and manipulation by insider Wall Street bankers. Factors that affect the price of crude on the open market are not the only ones in play. When hurricane Katrina devastated New Orleans and much of the Gulf Coast, the refineries there were severely damaged. At the same time, damages to offshore drilling rigs cut into our domestic supply of crude. This resulted in unleaded regular gasoline rising in price to more than three dollars a gallon at the pump. The combination of a year without damaging hurricanes with a new monster producing well in the Gulf of Mexico, brought those same prices down to just a few pennies more than two dollars a gallon. Since then prices have rebounded and roared past those previous highs with no end in sight.

Reduction in available supplies of unleaded gasoline caused by seasonal changeovers, refinery shutdowns, even increases in demand caused by more people traveling can cause gasoline prices to rise even as crude prices are dropping. As a result, many cried that we need to build more refineries—the news was filled with those kinds of comments. Then within a few months, the hubbub died down and wasn't heard again until the most recent drastic rise. This illustrates how human nature seems to drive us to put off doing things until after a disaster, even in our personal lives. That is especially true when politicians are involved. Locking the barn door after the horse is stolen is a major human trait.

Case in point on a much larger scale is some information about the hurricane Katrina disaster that the mainstream media is unlikely to report. While doing contract engineering for the U.S. Navy in the Philippines, I met an engineer who used to work for the state of

Louisiana. He explained that after more than ten years on the job, he became so disgusted with the situation in New Orleans that he quit the state and took a job with a private engineering firm. For years, he had submitted detailed plans for reinforcing the dikes that kept the Mississippi and Lake Pontchartrain out of the city. His plans were made in response to warnings of the Army Corps of Engineers and many others. None of his recommendations were ever followed. He said the politicians would approve minimum patch jobs on the most dangerous sections, sometimes just hours before a Mississippi flood hit. He predicted that any direct hit on New Orleans by even a weak hurricane would breach those dikes and flood the city. His plans would have doubled the thickness of the dikes and raised the roadway on the tops by several feet. The cost would be high, but only a small fraction of the cost that would occur should the dikes fail. A real prophet, wasn't he?

He also said, "Just how could any emergency vehicles move with most of the city under six to twelve feet of water." That was back in 1981! Fixing those dikes then would have been far cheaper than fixing them now and minuscule compared with the losses we saw from Katrina. Preventive maintenance is far cheaper than repair in virtually every situation—unless one is a politician! Those same politicians kept throwing the dice with thousands of poor people now paying the price. Of course, evidence exists that at least some money that did not go into strengthening the dikes found its way into the freezer of one local Democrat politician. Is he still in Congress? Hmm! It's not just gasoline that costs us money. How about those tax dollars wasted or stolen by corrupt politicians?

It is almost guaranteed to be the same with gas prices continuing to rise, particularly if we wait for politicians to do anything. And just whom do these politicians blame for high gas prices? Of course, they blame the oil companies, the oil cartels, politicians they do not like, usually everyone but themselves. Just how high will gasoline have to go before people demand viable alternates? Who will they blame for the time it takes to ramp up an alternate fuel system. It will certainly take years to accomplish during which time our economy could well collapse.

The most likely scenario that could save us from economic disaster is that an entrepreneur or group of them will see the possibilities described in this book or something similar, and build a new system and economy that will provide the needed energy. To some extent this is already happening as described in the appendix. As this is written, ethanol and bio-diesel plants are springing up all over the Corn Belt. Also, hybrids and *flex-fuel* vehicles are being provided in increasing numbers by auto manufacturers. American free-enterprise business will probably solve these problems long before government elites even get a good start. That is if these entrepreneurs can get going in the right direction before the government applies too many controls and wheels on those huge red tape machines.

Look at the Human Genome Project. It was completed in a short time by private research after it languished for years under government effort. It was completed in less than half the time predicted by the government and at less than half the cost, by private ingenuity paid for with private funds.

If those politicians, who regularly practice class warfare and hatred for business, would try to work with the creative genius of our American free-enterprise companies instead of doing everything to block their progress and success, many of these major problems would be nipped in the bud. That includes the human vector in environmental problems and even possible global warming. It is no wonder the Chinese are now making such rapid strides economically—they finally get it.

What truly seems strange is the emphasis on what is probably the most costly and difficult alternative energy strategy, hydrogen. Why the present level of government and media support for the hydrogen fuel cell vehicle is so strong, is baffling. This phenomenon demonstrates the power of so much misinformation about the long-range practicality and economic feasibility of this technology combined with the glamour of such a *magical* system—a system *that exhausts only pure water*. They are dead wrong about this. The energy balance of generating raw hydrogen by present technology alone will dump more carbon dioxide into the atmosphere than is currently being created by an equal amount of useable petroleum energy at the vehicle wheels. Future development of geothermal power to charge new and more efficient batteries, and new technology to make biofuels from waste, would not require the enormous investment in new distribution and storage infrastructure that use of hydrogen does. It could also be done much quicker, cheaper, and better in so many ways. Fortunately, the support for the hydrogen fuel cell vehicle seems to be waning rapidly as more realistic solutions are coming to the fore.

The Hydrogen-powered Vehicle, Is it a Scam?

Is it possible the hydrogen fuel cell system was being foisted off on the unsuspecting public by people who knew it would never be economically feasible simply *because* they knew it would never be economically feasible? This could have been orchestrated for the purpose of keeping other technologies that **are** economically viable from getting support, financing, or publicity. Considering the hydrogen economy advocates generously, they have the same kinds of views of the future as the people who tried to stop Eli Whitney's cotton gin and fought against automation in our factories. Again, these people are generally not evil, but merely people exercising that powerful survival instinct we all have. Just ask those who hold the patents on nickel-metal hydride battery technology why they are keeping this technology from being used in vehicles. Currently the cheapest and most reliable battery system for use in EVs and PHEVs, NiMH batteries have already proven economical, long-lived, and less costly than other competing technologies. Ask GM and Toyota why they are not using them.

This is but one among several sinister motivations that could be at work here. It includes serious efforts to stifle competition by one of the largest and most powerful groups of multinational corporations in league with many of the world's most despotic regimes. These people love the ruling power of petroleum. They will do virtually anything to thwart the efforts of any serious competitor. This is particularly true for any competitive system or organization with a real chance to capture the energy market with either a battery system or a usable liquid fuel and fueling system that costs less than gasoline or diesel. Many see GM and Toyota's efforts to take back and destroy all of the EVs they produced in the '90s as part of this alleged sinister plot. Could the Chinese, who ignore the restraints of our patent system, not now be developing if not building NiMH batteries for their promised electric vehicles?

The hydrogen fuel cell promotion, if successful, could waste billions of taxpayers' dollars and many years of effort on a program doomed to failure. This could easily delay other real, practical and affordable solutions. This book explains many of these alternatives in detail. A single caveat to this comment exists. Practical onboard generation of hydrogen could make the hydrogen fuel cell vehicle a possibility. Just such a system is presently under study and development at Purdue University. This system uses aluminum as an energy transfer from power plant to vehicle in a unique method developed at Purdue by electrical engineering professor Jerry Woodall. The system is described in section I-C-1 and the appendix in this book.

Fortunately, many real innovators see the immediate and pressing need for a simpler, more workable solution, and the potential of a number of currently available and emerging technologies. Many entrepreneurs are working diligently to bring to market products and technologies, that provide contemporary solutions to move us away from petroleum-based fuels. When they succeed, we will move into a more environmentally friendly energy system long before the hydrogen fuel cell system could hope to become a practicality. Many of these companies, their products, and systems, are described in the appendix at the end of this book. One danger we face is the orchestrated lowering of the price of petroleum to make petroleum fuels cheaper than others. Such an action by OPEC could jeopardize any program to develop alternative fuels and energy systems. One way we could counter this tactic is to place an import duty on oil imported from OPEC nations. This is outlined on page 111.

IX. A Bit of Speculation

This is the speculative section of the book where no holds are barred—anything goes, but it must at least have some basis in realism. It is where we brain storm ideas and work through them to understand their value and examine their possibilities. The realities and practicalities can be ironed out later. Many different existing, developing, and proposed systems are presented and described along with some hypothetical parts of our complete energy system.

Hopefully, those who make things happen—the movers and shakers among us who can take an idea and turn it into reality—will notice and advance some of the concepts in this book. This section is deliberately organized loosely to give creative minds the opportunity to wander among the ideas freely. Favoritism has been avoided and objectivity emphasized so as not to show any preferences. The obvious will still show through a bit. The ideas that led to this book started with the hydrogen fuel-cell system. At first glance, this seemed a marvelous new idea, possibly the first truly innovative vehicle powering system proposed in many decades. The idea of vehicles exhausting only water vapor was very attractive. In fact, it was so glamorous it caught the attention of politicians and the media. Such a radical and breathtaking new idea was easy for them to romance.

A closer look at the entire structure of any hydrogen economy shows it will probably be a very long time before it could become a reality, if ever, especially an economic one. Searching carefully the broad view of the energy industry, one finds that many other systems show promise of more rapid and practical conversion at much lower real costs. In the process of collecting this information, the author corresponded with a number of astute citizens. These included scientists, engineers, writers, workers, and professionals in many fields as well as others who are concerned. Nearly all raise the same question about each system discussed, including the hydrogen economy: *Is it truly a viable system?* A quotation from an article by *New York Times* reporter, Matthew L. Wald, sums up many opinions expressed about the hydrogen fuel cell promise in a few words:

"Despite the technological and infrastructure obstacles, a hydrogen economy may be coming. If it is, it will more likely resemble the perfume economy, a market where quantities are so small that unit prices do not matter. It will appear in items like cellular phones and laptop computers." [3]

The word ***evolution*** is used in descriptions of the necessary changes that will accompany any transition of such a massive and deeply entrenched system. It will come gradually and almost imperceptibly. Any changeover will be within the continuing evolution of an ancient energy system and a petroleum-based transportation system more than a hundred years in the making.

The petroleum industry and its companion, the auto industry, are old industries with firmly entrenched people and ideas difficult to change. Like every old, established industry, they have changed dramatically over the years, including much movement out of the United States. It is the pattern, of these old industries with their deeply entrenched and inbred managements, that most major innovations, especially those that have improved the industry and its products, do not come from within the industry. Nearly every major change has been introduced, sometimes with pressure, by people from outside the industry. The reason can probably be seen as an improper application of the old adage, *if it ain't broke, don't fix it.* Change is rarely accomplished from within these immoveable old giants because change is distressing and usually causes disruptions to highly entrenched people and technologies, not to mention powerful political ties. These industries certainly fit this pattern and will fight these changes. Unfortunately, with human nature being as it is, it will probably take a major catastrophe as described in the fictional news report at the beginning of this book to cause the public to support such a changeover seriously and overpower the political resistance to any major changes.

Hybrid cars are appearing on the road in steadily increasing numbers. Now it looks as though EVs and PHEVs are beginning to appear. A lithium ion battery aftermarket conversion for the Prius is available now, and other, similar projects are in development. GM's Volt is scheduled for sale to the public in 2010 and others like it will soon follow. What we now need most is a little rearrangement of priorities of some in the government and some in the industry. The main driving force of these changes will be a combination of public demands and economics. It is doubtful even powerful political forces will have more than a tiny effect on the coming changes. Those in government who have little real knowledge, will be funding endless and expensive research into systems they favor. They will also write endless laws to change and control. While they do so with great hullabaloo and endless posturing, free Americans, with their industry and ingenuity, and reacting to the real world, will quietly go about making these much simpler and more effective advances on their own.

The real purpose of this section, and the entire book, is to present many different ideas about the generation, transport and use of energy. The study of these ideas and the efforts to make them into realities can result in excellent and viable solutions in years, instead of decades. Creative solutions are sure to be found that require few and inexpensive infrastructure changes and by using both new and existing technologies. In Section III,

Energy Systems, Old, New, and Future, information about energy and vehicles describes the components that make up complete systems. By combining this information with existing technologies and many off-the-shelf components, we can find new and better ways of supplying, storing, and using power in the near future.

Any solution or group of solutions will be based on total energy systems. These include the creation, storage, distribution, and use of energy. The systems involved include power-grid stations, transmission lines, fuel procurement and manufacture, waste disposal, local power generators, vehicles and vehicle power systems, transportation and distribution systems for fuels, and maintenance and repair facilities.

In the end, these systems will do the following economically:

1. Replace fossil fuels, petroleum and coal, for transportation and electrical power.

2. Stem the drastic hemorrhage of money now going mostly to totalitarian nations that vow our destruction.

3. Provide an immense growth in our economy with many high-paying jobs right here in the United States. It could easily be the biggest opportunity for growth in our economy ever and a real answer to many environmental concerns.

4. Stop or at least slow the rate of addition of carbon dioxide in our atmosphere.

A PHEV is any vehicle with onboard energy storage that can be recharged by connecting a plug to an electrical power source or by using an onboard generator set—an engine-powered generator. This gives PHEVs the characteristics of both conventional hybrid electric vehicles and of EVs or battery electric vehicles. Among possible PHEVs are all types of vehicles including passenger vehicles, commercial passenger vans, utility trucks, school buses, RVs, and military vehicles.

The operating costs of these vehicles based on the cost of electricity used to recharge their batteries are around a quarter of the cost of gasoline used in the generator set at current electric rates and fuel prices. If PHEVs use nonfossil fuel for on-the-fly charging, no net carbon dioxide would be added to the atmosphere. If, during their all-electric operation, their batteries are charged from sources such as wind power, hydro power, nuclear energy, or geothermal, no net carbon dioxide would be released. Such use can help us reduce: the increasingly expensive dependence on petroleum products. It will also help satisfy the global warming crowd by producing less carbon dioxide than fossil fuel-powered vehicles. Other potential benefits include the following: improved national energy security, less frequent refueling stops, easy plugin recharging at home overnight, and the possibility of providing emergency backup power in the home.

As this is written, no plugin hybrid electrical passenger vehicles are being mass produced. Several EV and PHEV conversion kits are now available including one to convert Toyota's Prius to a PHEV with a lithium ion battery pack. Toyota and General Motors have announced plans to introduce production PHEVs, GM's Volt introduction is promised for 2010. Conversions of production model hybrid vehicles are available from conversion kits and conversion services. Some existing PHEVs are conversions of Toyota Prius hybrid cars with battery packs that extend electric-only range and add plugin charging capability.

Current hopes for PHEVs with all-electric ranges of several hundred miles are dependent on the success of several new battery technologies announced recently that are just emerging into the marketplace. Information about these new batteries is covered in the appendix at the end of this book.

For several overwhelming reasons, we must quickly develop innovative vehicles that run on alternative systems and/or nonfossil fuels. The need to develop and build the infrastructure to create and distribute those fuels will accompany any new system.

The first is the menace of decreasing supplies of petroleum-based fuels amplified by the threatening situation in most of the oil-producing nations. This combination threatens to remove rapidly increasing amounts of money from our economy and put it in the hands of those who would destroy us.

The second is the increasing damage to our economy caused by both the rising costs of energy and the movement of vast sums of money from our economy into those of the oil-rich nations.

The third is the power of the global warming or climate change adherents. Whether this supposed threat is real or imaginary is irrelevant as the movement itself is still a real menace. The efforts of so many to convince the public that it is a reality, has created an active *global warming* movement that wields a great deal of power. This movement also extracts a monumental cost from the world's economy. Disarming this movement by removing the root cause for its existence will certainly save energy, money, and a lot of grief.

Conversion to nonfossil fuels will answer each of these threats by quickly developing a new American energy/fuel industry. Such an industry would substantially expand our economy and stop the increasing percentage of carbon dioxide in the atmosphere. Even those who disagree about the threat of global warming must agree that stopping the rapid hemorrhage of cash out of the United States and into the Middle East and other oil-producing nations that vow our destruction is a major priority. That cash outflow may spell disaster for our economy if it is not stopped soon.

The current commitment to building numerous ethanol and biodiesel plants throughout the country is evidence that we are at least starting toward the goal of energy independence.

Here's the only real question. Can we convert to alternative fuels or energy systems fast enough to avert the menaces mentioned in the previous paragraphs? Ethanol alone is not the answer for several reasons.

1. It may take more agricultural production capability than we have available to produce enough ethanol to fill the demand.

2. Other fuels may be cheaper and better suited to certain types of vehicles.

3. As now produced, ethanol consumes a great deal of energy and releases carbon dioxide into the atmosphere in its manufacture.

4. Use of food crops to produce fuel is already raising world food prices, a situation that can only get worse as fuel prices rise and demand for the raw material for fuel increases. Putting all of our (fuel) eggs in one basket may not be such a good answer. The development of other nonfossil fuels that can be blended with ethanol or used in different types of engines can work to make the change quicker and not overload any sector of our economy. Given the chance, economics will determine what vehicle types, energy, fuels, or mixtures of fuels will best serve our overall needs. It is unlikely the hydrogen fuel system will become practical or economically feasible for use in private vehicles within the time available and probably never.

Flashy new cars are the most visible and glamorized part of the energy use system. The basic design of the motive force that drives these new vehicles is almost totally dependent on the energy system used to power them. Like research into and development of the hydrogen fuel cell vehicle, this is about a completely new and different energy system of which the vehicle is but a small part. The current transportation energy system includes oil exploration, drilling, extraction, storage, refining, transportation and distribution, and the development, manufacture, distribution, repair, and maintenance of vehicles. Any vehicle would be but one part of any new and different system, designed from scratch, manufactured, and put in place safely and economically. This is a daunting and terribly expensive undertaking, which will take a long time to complete and will have many unforeseen pitfalls along the way.

So we must consider alternative systems based on nonfossil fuels or other energy systems that would have advantages over petroleum-based energy. We call some of these fuels *biofuels* for their biological sources. Biofuels include methanol, ethanol, butanol, biodiesel, and any other fuels made from renewable, biological sources such as crops and agricultural waste. These fuels would use existing, proven technologies and components including existing systems of storage, transport and distribution with little if any modifications. Existing vehicles and small engines could be easily and inexpensively converted to use biofuels. Addition of biofuels to service stations would be no more challenging than the addition of diesel fuel pumps and tanks accomplished without fanfare a few years ago. The required changes would evolve with little disruption to our current system.

Unfortunately, biofuels produced from normal food crops alone may not provide sufficient quantities of useable liquid fuels to meet demands. Their use could interfere substantially with world food production and prices. Other systems could provide huge quantities of biofuel at quite low prices and still not affect world food supplies. **At the least, It is likely that if properly implemented, the best combination of these new systems, biofuels and battery powered vehicles, would remove our dependence on foreign oil within ten years or less using sound economics.** Couple this with both the new engine technology that can use a variety of liquid and gaseous fuel, and promising new battery systems that will provide portable power, and our future without dependence on fossil fuels, and especially petroleum-based fuels, is assured. This comes with many benefits including both economic and environmental, while at the same time, calming fears of global warming.

Should we adopt the best program from among the available options, the benefits to our nation and the world would be substantial and almost immediate. What can be called the Optimal Energy Economy, is that combination of systems using biofuels and/or battery power that would be far superior to both the existing petroleum-based system and the proposed hydrogen fuel cell system. Its environmental effect would be minimal. It would require much less energy to implement and could be developed so no fossil fuels would be used. **Use of any fossil fuel adds carbon dioxide to the atmosphere and removes oxygen.**

The financial boon to our nation would be tremendous as all our money now used to buy imported oil would go for American nonfossil fuels or electricity. With existing infrastructure used for nonfossil fuels and expanded for the distribution of electric energy no huge outlay for entirely new systems would be required as with some systems. Fuels produced without using fossil fuels would be much less costly than hydrogen at the pump. The least costly process for producing useable biofuels would shake out from among those we now use and will develop in the future. Among the promising ones are several using biological agents including micro biota and algae that could make useable fuels from waste water and agricultural waste materials. This could reduce the pressure on grain supplies that threaten to disrupt the food-chain and raise food prices. It makes no sense to save money on fuel and then have food costs consume more than the savings. Ethanol and biodiesel are already being produced from renewable plant waste in small quantities. What we need to do is find more efficient methods to use waste materials rather than food material.

Engineers are mostly pragmatic-thinking individuals, who find, develop, and implement ways to create useable products and services using technology. They do so by digging or pumping material out of the ground, pumping it out of lakes, rivers, and oceans, compressing it out of the atmosphere or manufacturing it from natural vegetation or crops from the field. Sometimes these processes damage the environment, but mostly the damages are corrected with a few nasty exceptions caused by some thoughtless or ignorant people. Engineers must live in the same environment as the rest of humanity so they would like it at least as clean and livable as anyone else. This should be one of our guiding principals in any solution to the energy crisis.

Even without considering possible global warming, we desperately need an alternative to petroleum products that are becoming more difficult and expensive to find and recover. A sudden, major disruption of the oil supply would wreak havoc with the world economy and create a depression that would make the one in the thirties look mild in comparison. If and when the Middle East implodes in terrorism, we will need to have such an alternative in place, or our economy will suffer possibly irreparable damage. The systems described in these pages could provide part or all of that alternative.

It is interesting that the rapidly expanding economies of India, China, and other third world nations have created a rapidly growing demand for petroleum and other fossil fuels. They will continue to do so for years into the future. China is currently on a binge of building coal-fired power plants. These power plants are coming online as often as one a week as this is written. Consequently, China passed the United States in carbon dioxide emissions to the atmosphere during 2007. Can India be far behind? For economic reasons, the Chinese or Indians may be the first to adopt the systems described in this book and thus rapidly outpace the rest of the world in cheap energy production. Cheap energy is now and has been for some time the single most powerful positive economic force in the world. China could soon emerge as a major beneficiary of alternative energy, far outstripping the United States and the rest of the world. This would only be possible if they become energy self-sufficient. Alternative energy systems, possibly developed from information outlined in this book, will be the key to economic superiority and domination of the world.

X.   Putting it All Together

The Optimal Energy Economy

The Optimal Energy Economy is a descriptive name for what is the main subject of this book. It is the economy that would result from the adoption of the most efficient combination of the systems described herein and other systems yet to be designed and described. The name comes from what is undoubtedly the heart of any successful system, that being the optimal use and conversion of energy. No matter how one looks at it, the entire system is all about energy, its capture, use, and distribution. Energy, like matter, cannot be created or destroyed. It can only be converted from one state or form into another and always with some losses.

The Optimal Energy Economy consists of many viable systems to replace fossil fuels with renewable fuels that add no net carbon dioxide to the atmosphere. In the end, it will be cheaper, quicker, and far more practical than any other system. The hydrogen fuel cell vehicle with its expensive new infrastructure requirements, is just one system among many that may not be very practical. Optimal Energy Economy systems will use existing and newly developed technologies in easily adaptable solutions. Parts of the system are available right now and those that are not are distinctly within our current technical expertise. Certainly it could be carried out in just a few years. Combine this with creative new developments and we could thumb our collective noses at the world's oil-supported despots.

This book proposes timely, affordable, and practical parts of this economy, solutions to the energy crisis facing our nation including:

1. The growing consumption and dwindling supplies of petroleum-based fuels are seriously affecting the entire world. This is aggravated by the now rapidly expanding economies of the nations of China and India among many others.

2. The rapidly rising cost of foodstuffs caused by a combination of two significant factors. One is the conversion of grains from the food industry to the fuel industry. The other is the increasing demands of the people in many nations like China and India that are now expanding their economies.

3. The growing amounts of carbon dioxide in the atmosphere and its possible link to global warming.

4. The increasingly dangerous transfer of tremendous amounts of money from the United States (and many other free world countries) to totalitarian regimes that threaten, indeed promise, our destruction.

Recap of Energy

All of our electrical energy is generated in power plants of some kind, most commonly by heat energy. Even nuclear energy is the conversion of mass, a form of captured energy, into free or useable energy as heat. In modern nuclear power plants heat from fission is first transferred to liquid sodium or another high-temperature medium that carries it through high temperature pipes to a water boiler where it generates high pressure steam. From this point on a nuclear power plant operates precisely the same as any other heat energy power plant. The steam goes through a turbine connected to an electric generator. The turbine generator converts the heat from the steam into electricity transported into the grid that distributes the energy to end users from factories and municipalities to homes and offices.

Energy that starts as nuclear is first converted to heat in steam and then into mechanical energy by turbines and finally into electrical energy by generators. Electrical energy is then distributed by the electric grid and then converted into light and mechanical energy in tools. This includes everything from arc furnaces for steel to kitchen mechanical mixers, to office computers, to the charger in a few new battery-powered electrical vehicles.

Coal-fired power plants currently supply more than half the energy used in the U.S. In these plants, heat is supplied by the combustion of coal but the turbines and generators are virtually identical with those in nuclear power plants. The advantages of nuclear over coal-fired boilers include lower cost, higher safety, lower pollution, and no carbon dioxide emission. Expensive safety measures raise the costs of the capital required to build and maintain a nuclear power plant to the point where it costs about the same as coal-fired energy.

Whether it is to light a city, power a factory, move a family car, pump irrigation water, or heat a home, it is all energy. Whether it comes from an electric outlet, a tank of fuel, a trainload of coal, a nuclear reactor, a battery or group of batteries, a fuel cell, a solar cell, a hydroelectric turbine, or a wind-driven generator, it is all energy. Whether measured in kilowatts or horsepower, it is all energy. Energy is what we all pay for one way or another. Generally, the more power we use, the more we pay. With gasoline costs varying between two and four dollars a gallon, most people pay ten to twenty cents per mile for liquid fuel. The same amount of energy from your household outlet would cost about four cents or a quarter to half the cost of liquid fuel. The problem is to get energy from the electric generator into a practical vehicle, economically and efficiently, and to give that vehicle sufficient range to satisfy those who use the vehicles. In effect, that is the real point of this book: to examine as many ways as possible of providing portable energy in a practical way and at reasonable cost to the traveling public.

Geothermal Power Is Possibly the Best Solution

There are many possible ways to power mobile vehicles other than using almost exclusively gasoline or diesel-powered internal combustion engines. Using geothermal power to generate electricity and developing a practical method of putting that energy into a vehicle that can use it seems to this writer to be one of the best overall solutions. This combination could be the least expensive alternative to the present system in the long run. It could also be the best for the future of the environment. What remains to be discovered is whether it will involve hydrogen and fuel cells, innovative new batteries and/or fuels, or technologies yet to be invented. Why then is a solution so imperative?

We are running out of cheap oil that currently adds large amounts of carbon dioxide to the atmosphere. Coal-fired power plants also add massive amounts of carbon dioxide to the atmosphere. Use of crop-based fuels would negatively affect the food supply. Other sources of electric power all have serious drawbacks. Though nuclear, geothermal, and wave action energy are the safest, least intrusive, and least expensive sources of energy apt to be available in the foreseeable future, nuclear is a red flag to many people. Because of this, we have not built a new nuclear power plant in America in more than thirty years and there are none proposed at the present. That leaves geothermal and wave energy as the most logical choices for energy expansion.

Though but a tiny part of our current energy mix, geothermal looks to have the greatest potential and the least number of negatives. Because of the minuscule portion geothermal energy makes up of our total energy mix, it is the least understood and will therefore require the largest amount of new technology development to become a significant contributor. This must be done at reasonable cost to satisfy our growing need. Useable geothermal energy underlies about 60 percent of the continental United States. A geothermal power plant costs about the same as a coal-fired plant. These are facts that should grab everyone's attention. Geothermal is probably by far the most economical, the safest, and the least environmentally damaging power generating system we could use in the foreseeable future. It is not seen herein as a possibility, but as the best probability—maybe even the only one.

The other needed component of the transportation sector of the Optimal Energy Economy is power for vehicles. Whether it is a pure electric vehicle or a plugin hybrid electric vehicle, both are described earlier in detail. The combination of these two factors and their importance is really about energy, the economy, the environment, and what we must do about it. During any changeover, nonpetroleum fuel will increasingly be needed for existing vehicles. Whether it is ethanol, butanol, biodiesel, or other fuel or blend of fuels, the need is there. With gasoline and other petroleum-based fuel prices jumping almost on a daily basis, these alternative fuels will become increasingly attractive. Economic pressures alone should generate frenzied efforts to get these energy products to market quickly along with new technology vehicles.

This book is directed toward what is unquestionably the most pressing issue facing our nation—one that bears on or affects virtually every other problem we face. Our economy is tied directly to the cost of petroleum-based energy as is the security of our nation. If we do not do something about this immediately, we will be devastated economically while the tyrants and despots of the petroleum-exporting nations will succeed in their avowed efforts to destroy us. Our mainstream media could be a powerful positive force in this desperately needed effort. If only they would quit concentrating on political activism and sensationalism, they could be a tremendous help. They could leave that to the politicians and tabloids and start pressing for actions like those described in this book.

Scientific American reports on both geothermal power and electric vehicles in their July 2010 issue. Both articles pose pro and con arguments. The article by Jane Braxton Little is about the 200 megawatt Calpine power plant that supplies electricity to Santa Rosa, California. It is near a volcanically active area in the Mayacama Mountains. The article describes a situation where waste water is injected into hot rocks above a mile and a half deep aquifer that rests atop a deeper magma chamber. This is certainly a win-win situation for the area residents. The few more small earthquakes are a minor negative. In the article, Ms. Little says, "By generating 200 megawatts of electricity from wastewater, Santa Rosa and Lake County have effectively reduced greenhouse gas emissions by two billion pounds a year. The city and area towns have also stopped pouring effluent into the Russian River and Clear Lake and have eliminated the need to build new storage and treatment facilities."

There is a great need for more extensive research into deep-well geothermal as it could be available and economically feasible in a majority of areas all over the US. Considering the costs of a similar sized nuclear facility and all the public resistance to such installations, geothermal could be our best environmental and economic bet for the long-range future.

Ms. Little's closing comment: "For the many potential sites that lack an adequate supply of water to inject into hot rocks, the power plants at the Geysers still serve as an inspiration. They have demonstrated that treated effluent is a commercially viable alternative to fresh water for steam-generated electricity."

Santa Rosa official, Dan Carlson says. "Of course, safety issues require more study." But he is optimistic: "Our residents are benefitting, the environment is benefitting, and people all over the world can use this model to improve their own communities."

The other article about electric vehicles seems to make the case that plug-in electric power generates more CO_2 at the power plant than gasoline-powered cars do. The author, Michael Moyer, says, "On regions powered mostly by coal—a much dirtier fuel—electric vehicles will lead to an increase in the amount of carbon dioxide released into the atmosphere." I find this comment is in stark contrast to everything I have read and studied about the efficiency of power plants compared with gasoline or even diesel engines. It would also mean that the darling of environmentalists, the hydrogen fuel-cell car, (not mentioned in the article) would also generate more CO_2 than a similar gasoline powered car.

It appears to me the article and its title, *The Dirty Truth about Plug-in Hybrids*, is written using misleading implications, especially in the comments about the *zero-emission tour*. No, EVs in our present energy mix could possibly be zero-emission vehicles. If I read his graphs correctly (and they are unclear) the CO_2 emissions of both EVs and PHEVs are less than ordinary hybrids in every graph shown. That would also make their CO_2 emissions much less than that of nonhybrid, fueled cars.

There is another article in the same issue about what they claim is a controversial technique to recover natural gas from extensive deposits of shale. The technique is called *fracking* and involves drilling horizontally into the relatively thin deposits of shale and using an extremely high pressure mixture of water and sand with about 0.5% chemicals to fracture the shale and thus release the methane which comes up the pipe under pressure. The chemicals serve many purposes including, lubrication, prevention of rust, cleaning the pipe perforations, and others.

This is the third article in one single copy of *Scientific American* addressing some of the same energy related factor addressed in this book. Energy is probably the single most powerful factor on our economy. If we have a plentiful supply of energy at relatively low cost, our economy will boom. If it is expensive and in short supply, our economy will tank. This is true of the entire world as well as the US.

Some possible serious side effects of a major reduction in our use of petroleum

Shut off the billions of petrodollars now feeding the tyrant regimes of Iran, Saudi Arabia, Venezuela, and even Russia, and they would have to reform or die. Stopping this financial hemorrhage would also add those billions to our own economy giving it a huge boost. Each dollar kept here and not sent overseas actually increases our economy by two dollars: the dollar we did not spend and so remove from our economy and the one we created or added to our economy. The billions of new dollars created right here and added to our economy would provide many high-paying jobs for Americans.

We must use caution about oil pricing and OPEC. When our investment in new fuel industries begins to affect the market for petroleum products, the oil-rich nations could drop the price of petroleum enough to affect the profitability of alternate fuels. Should this happen just as new manufacturing facilities and technologies are coming on line, it could be devastating to those industries and to our economy. Lots of oil is still available so this is a definite possibility. We could take steps to prevent such a disaster. Our government could set a floor price for imported petroleum at say $50.00 a barrel. Domestic petroleum, and that from Canada and Mexico, would be free of this tax. Unlike setting a maximum price that just does not work, a minimum price would work perfectly. Suppose the world price of oil dropped to $30 a barrel. Every barrel imported would have a $20.00 tax or duty added making the price $50.00. The free trade people would scream, but such a tax would stabilize our price of oil and maybe even that of the world. This would protect our fledgling alternative fuel industry and encourage the advancement of new technology power and vehicles.

Environmentally, the changes proposed would quickly end or at least slow America's part in the growth of carbon dioxide in the atmosphere. Whether or not carbon dioxide is a significant factor in global warming matters little to the importance of these solutions since the result of their adoption would still provide positive change. Many other factors are involved in our economic success, but solving the energy crisis would be major among them. The steady increase in the cost of energy to power our vehicles, factories, and homes, is already strangling our economy. This is becoming evident to all who drive past those signs posted in every filling station in America. Without an assured supply of low-cost energy, our economy and our people are facing a growing threat and will suffer accordingly.

Another big, inspirational program and the leadership to start it is essential to the successful solution to the energy crisis. We should be hearing the same passion in our leaders as John Kennedy had in responding to Sputnik by committing us to putting a man on the moon in ten years. We need a ten-year commitment to replace petroleum as the source of our energy. This challenge is equally important—maybe even more so. The scope of the project is even larger—far larger. Its direct rewards promise to be infinitely larger in every respect and especially regarding our economy and leadership position in the world. It is easily within our power to start right now to do so. It is virtually certain that whether we do it or not, one or all of the nations of China, India, Japan, South Korea along with many others, will. Should that happen, the United States will probably be left in the dust economically and intellectually. In actuality, the French have already taken several important steps and are well on their way to achieving a major part of this proposed system, even if they fail to realize it. Their expansion of nuclear power is already growing their economy. With increasing sales of electricity to other members of the European Union they are not only boosting their own economy, but by providing inexpensive energy to the European Union they are boosting the economies of all member states. This is not going to happen in the future. It is already a reality.

XI. Wish List—
Things We Wish Were Available Now

The author puts at the top of the list, one single powerful action our legislators could take that would quickly bring about the desperately needed changes in our energy systems. That is to remove all taxes on profits and capital gains from that part of any company or individual directly involved in developing and producing any new component or system that takes us away from fossil fuels and petroleum in particular. This tax moratorium could be for a period of ten years from the enactment of the law or possibly ten years from the startup of the development. This would provide a powerful new incentive for entrepreneurs, developers and investors to go into the alternative energy business. It might even tempt the oil companies to invest some of their profits in alternative energy.

It is a proven fact that high business taxes drive business from our shores even as they satisfy the warped class hatred fomented by those on the left. It should be apparent that lack of taxes will encourage business development where we need it and provide many good paying jobs in the process.

Several specific items or systems desperately and quickly needed are dependent on the tax moratorium at least for a powerful incentive.

A. A PHEV or EV Conversion for Existing Vehicles

Hopefully, one or several entrepreneurs will develop and produce one significant item. That is a package conversion for current gasoline powered vehicles to either a pure EV or even a PHEV. It could be a combination of a battery system and electric motor that would replace the engine/transmission system of present front wheel drive vehicles. A similar package especially designed for trucks and RVs is also desirable.

The math works out this way. At present prices, 20,000 miles represents an expense of roughly $4,000 for fuel at $4.00 per gallon consumed at 20 miles per gallon. Electricity at a nickel a mile translates to only $1,000 for 20,000 miles. That is a savings of at least $3,000 a year for the average motorist. Even at twice that mileage rate or 40 miles per gallon the savings would be about $1,500 per year or $125 per month.

Consider this: AC Propulsion is producing and selling just such a conversion for the Scion xBox. The conversion itself costs $55,000. Add to that the $15,000 cost of a new xBox and the total vehicle price is around $70,000. For more details about the eBox, go to the AC Propulsion website:

http://www.acpropulsion.com/ebox

The factory xBox averages around 30 miles per gallon making fuel cost around $2,667 for 20,000 miles. Electricity for the eBox costs about a nickel per mile. This means a savings of $1,667 in 20,000 miles. Based on fuel savings alone the entire cost of the conversion would be recovered in 650,000 miles however long that took.

For the average motorist this would mean about a 3 percent return on investment, not a good deal. But wait a minute—there are other cost differences including routine maintenance, obsolescence and a few intangibles. Maintenance costs for electrical components are much less than for IC engines and transmissions and would be reflected in both out-of-pocket expenses and maintenance contracts. The savings would amount to between two and three cents per mile for the average motorist. Add to that the fact that EVs depreciate far less than normal vehicles and even with the addition of conversion costs, savings of nearly another dime per mile is likely. This all translates to cost recovery for the conversion at 260,000 miles or about a 7 percent return on the investment. Financing charges for any additional loan would eat into those amounts.

Though that may seem quite expensive, commercial fleets that put on 200,000 to 300,000 miles per vehicle per year, would get their money back in as little as nine months. That would make these vehicles very attractive to fleet and commercial users.

Supposing a similar package was developed to be installed in an existing vehicle. Suppose mass production brought the price of this package down to $10,000 for the average car. Would there be any buyers? Using the numbers listed above for the cost of gasoline or electricity and for the savings on maintenance and depreciation, 10,000 miles would cost $3,100 less. At this rate including adjustments for differences in maintenance costs and depreciation it would take a very acceptable three years to recover the cost of the conversion. Driving 20,000 miles per year, the monthly savings would amount to $220. The monthly payments on a five-year vehicle loan at 6% would be about $215. The savings would more than cover the monthly payment. If the owner drove twice that amount or 40,000 miles the savings would cover all of the monthly costs and put $225 in cash into the owner's pocket every month.

In the RV industry, a $60,000 conversion for class A motor homes would include a seventy-five-horsepower or larger, diesel powered generator. That sounds expensive, but these current units get between five and six miles per gallon yielding a cost between seventy and eighty cents per mile. With electricity, and much reduced diesel fuel costs of twelve to twenty cents per mile, and maintenance and depreciation costing so much less, the savings per mile could be between fifty and seventy-five cents, or $5,000 to $7,500 for 10,000 miles. That would mean the conversion would pay for itself in 80,000 to 120,000 miles. Add to that the value that a plugin hybrid electric RV (PHERV) would retain over its lifetime and the results would thrill many RV owners.

New products are always far more expensive when they first appear than later. The wealthiest among us will be the first customers and help the market develop. The price of computers has dropped precipitously from the simple monsters of the 1980's at ten thousand dollars and more to infinitely faster, smaller, better PCs at less than a thousand just ten years later. Since then, an equally fantastic jump in performance led to today's systems. It will be interesting to see what the selling price for the new EVs and PHEVs soon coming to market will be. It is conceivable that the combination of technological advances and mass production of EV components and batteries will drop the prices of these components to less than those of their IC counterparts in current vehicles. Electric motors and generators are far simpler than reciprocating engines and automatic transmissions and wear out much more slowly. It is quite conceivable that the cost of EV components could drop well below those of IC engines and transmissions in the future. Such being the case, the electric vehicle, whether PHEV or straight EV, looks to be a real winner in the future. Their success depends on a large increase in electric generating capacity and addition to the distribution system. Vehicles alone will not solve the situation.

Fortunately, people are working diligently fulfilling my wishes. Check out:

http://eaaev.org/eaalinks.html#EVconversions or

http://www.pluginamerica.org/links.shtm

Many other web sites are touting EVs, PHEVs, and other electric vehicles and conversions. Search them for more information.

B. Geothermal Energy

With coal-fired power plants spewing tons of carbon dioxide into the atmosphere I do not see their growth as a possibility whether or not their carbon dioxide contributes to global warming. Even if it does, it will probably not be the disaster the prophets of doom are predicting. Likewise, the opponents of nuclear power will probably defeat all efforts to develop this source of new power whether the danger is real or fabricated. At this time, the actual safety record of nuclear power worldwide is far better than any other energy system, but opponents ignore this fact and use scare tactics and emotional appeal to promote their agenda. Of course, our politicians in their infinite wisdom have established insurmountable roadblocks to new oil exploration and drilling while preventing the building of new refineries. That leaves us no practical economic alternative. At least at the present, no group has expressed organized emotional opposition to geothermal power. It does not cause pollution. It does not put carbon dioxide into the atmosphere. It does not pose any threat of radiation. It does not kill migrating birds. It does not destroy or modify the ecosystems in rivers or estuaries, and it does not use up any diminishing resources. Since the plant has a footprint similar to a nuclear plant, opposition is sure to be mounted because, *we do not want that eyesore in our neighborhood.* Fortunately, geothermal power is available directly beneath us everywhere and is currently economically reachable in more than 60 percent of the continental United States. It is described by energy experts as, *the most underutilized, easily available energy resource on the planet.*Perhaps those oil drillers can use their considerable expertise and experience to drill for heat. Converting their technology and equipment to tap the unlimited store of heat inside the planet should be comparatively easy. Likewise, the builders and assemblers of steam power plants could use their expertise to design and build the power plants to use the steam geothermal wells can tap.

C. Then There is Butanol

A wish for a new direct replacement for gasoline would also be high on the list. Maybe butanol is that answer. If it is, we'd better get production rolling and soon. The practical method to produce butanol from waste material would keep our old gas burners running. It could be pumped from the same pumps and into the same tanks and engines we've been filling with gasoline for a hundred years. Currently the Illinois Missouri Biotechnology Alliance estimates the cost of butanol made from hydrolyzed corn fiber at between $1.40 and $1.60 per gallon. That would make it price-competitive with unleaded gasoline that it could replace directly. Made from hydrolyzed corn fiber rather than feed corn it would not significantly impact the food section of our economy yet would add considerable value to our total corn crop. Other brand new ideas are out there being developed by entrepreneurs that have yet to garner enough attention to be noticed. Unfortunately, a lot of charlatans and con artists are also out there with convincing though false stories. Time and patient investigation will be needed to learn the difference between the real, the practical and the useless or fake. Even those with the best intentions can be so caught up in their dreams as to miss the obvious flaws that make it impossible. Often a lot of money gets spent before an impractical idea or even complete hoax is revealed. The hydrogen fuel cell vehicle is a classic example of one impractical but easily romanced system.

XII.   Conclusions and Predictions

Why We Are in This Dangerous Situation

As a nation, we are addicted to petroleum fuels almost like a heroin addict is addicted to that drug. Billions of dollars are being extracted from our economy, some to feed the appetite of despotic leaders of enemy nations. We continue to give these despots increasing percentages of our GDP in exchange for this diminishing resource. Fortunately, most of our imported oil comes from friendly nations like Canada and Mexico. The pollution damage to our environment, and carbon dioxide contribution to possible global warming, are other concerns. Since starting on the research for this book ten years ago, many of the things and systems described in these pages have become reality. The technology is developing and changing so rapidly it is difficult for anyone to keep up. I have no doubt there are products and systems not mentioned in this book that could quickly come to the fore. At least three technologies in this edition have come into existence in the last year or two. It is highly unlikely that environmental activists and opportunists will relinquish their hold on either nuclear power or most of our northern or offshore oil exploration. No matter how foolish, irrational or self-serving they may be, these ideas have become well developed in the psyche of the media. Members of the media have in turn imbedded them in the minds of the public who react emotionally about them even without reason. Developing alternative fuels and power sources may be simpler and less expensive. The global warming phenomenon is probably in the same category. Even in the worse case adherents have supposed, the effects of global warming are actually quite slow and benign. In addition, whether it is human activity and use of fossil fuels that are causing global warming is questionable in spite of its acceptance as proven fact by many and especially the media.

What I see as the best combination of energy systems for our future

I have been collecting information about the various energy systems for this book while writing and organizing it for almost ten years. During that time I have changed my mind several times about what I think is the best combination of systems to solve our energy problems. Presently, the combination I have labeled as the Optimal Energy Economy is the development of PHEVs and the expansion of our electric generating capacity with geothermal power plants. The optimum PHEV looks to be powered by a small diesel engine/generator combination running on biodiesel made principally from oil-producing algae grown on waste products. Since it uses no food crop in its manufacture, use of this oil would not raise food prices. This oil should be quite economical, require no petroleum, and will return that carbon dioxide to the atmosphere previously removed by the algae. It will therefore be a no net carbon dioxide fuel. I believe diesel PHEVs already exist, at least on the drawing boards of several auto manufacturers. Spark ignition engines running on butanol or DMF could also be designed to power PHEVs including the Chevrolet Volt scheduled for release in 2010. These PHEVs will have plugin charging for their lithium-ion, firefly, or future battery systems. These batteries will be capable of moving them more than a hundred miles on battery power alone.

An unusual conclusion

After all the research of information about energy and fuels used to write this book it is this author's opinion that there is one best possible solution for practical, affordable energy and its use. **I would urge those in power to consider doing whatever is required to make such a system a reality.** That total system involves electric vehicles, EVs, and plugin hybrid electric vehicles, PHEVs. These vehicles are powered mostly by electricity from batteries charged from an electric grid supplied by geothermal power.

GEOENERGY is the most abundant and widest spread source of heat on the planet, yet it is rarely addressed. It is virtually inexhaustible, economically available, nonpolluting, noncarbon-dioxide-emitting, and grossly underutilized. A geothermal power plant costs about the same as a coal-fired plant of the same capacity and has a smaller footprint. Once erected, no fuel system is required so maintenance is the only ongoing cost. It is potentially the least costly form of power generation available and certainly has the lowest environmental impact. It is an environmentalist's dream come true. Its use requires drilling for heat almost exactly like drilling for oil, a well-developed technology. Why so few people ever mention it is a mystery. The development of geothermal energy to replace retiring coal plants and provide the necessary increase in electric generating capacity could be the best way for our future. With technology that is presently just beginning to grow, improvements in cost and performance could easily make it the best and most economical domestic source of electric power. This would satisfy complaints of both the global warming and anti-nuclear crowd at least as far as generation of electric power is concerned.

The author suggests a novel engineering process for geothermal energy. The possible increase in minor earthquake activity is one of the drawbacks to geothermal power plants using water pumped directly into hot rocks as in the Santa Rosa power plant described on page 153. By using a closed system of one or more wells drilled directly into very hot rocks, this problem could be overcome. Once the hole is drilled and a closed liner pipe placed at the bottom, water could be pumped through a second pipe inside the casing down into the hot section. There it would pick up heat, flow up through the casing, and flash into high pressure steam at the surface driving conventional steam turbine generators. This would also solve the problem of corrosive steam from ground water damaging the turbines. A heat exchanger system using liquid sodium might even be practical. Just such a system has been developed and is used in nuclear power plants. All that is needed for this to become practical is a little engineering genius proving once more that those people saying "it can't be done" are so often passed by people "doing it." This new technology would provide excellent opportunities for engineering development along with many new high paying jobs, all right here in America.

A Practical Interim Fuel

Butanol is a practical interim fuel that could ease any major vehicle changeover. Successful creation of a butanol from corn fiber industry (and maybe even from other waste products) could create a fuel that would directly replace gasoline in virtually every way. This looks to be the best answer for fueling the millions of vehicles that now run strictly on gasoline. Not only would it replace a fossil fuel, but it would save money by not obsoleting all gasoline engines. That would save not only many vehicles, but an entire industry that has built those *gas buggies* for more than a hundred years. Like going to the moon, all it will take is dedication, hard work and support from the public. It could even be a rallying point to get our warring political factions to work together. Many benefits would accrue were we to work together on such a worthwhile endeavor rather than using hate speech and calling each other names. If we don't find ways to work together, this dream will be slaughtered on the field of political battles of rival ambitions.

Go Green on a Small Scale - Homes and Businesses

What can you do as an individual with your home or small business? The answer is there are many things you can do that will reduce your energy use and save, even earn, money. At the same time you will help reduce CO_2 emissions. From buying energy saving lights to revolutionary new insulation products to solar or wind energy installations, there are a great many new technologies and systems now on the market. These include both solar and wind systems for generating substantial amounts of electric power with small home installations.

Solar: Solar power systems run from lights to battery chargers to water heaters to complete power systems. You'll be amazed at what is now available for solar power, and how prices have dropped. Systems are available in many sizes and power outputs, even for RVs.

Wind: From wind turbines of many types to large and small towers, an explosion of systems suitable for the small to medium sized installations has come to market recently.

New systems are constantly being developed and put on the market in an explosion of revolutionary eco-friendly energy products.

IMPORTANT: Many of these items carry substantial rebates from both state and federal governments, some as high as 90%. You can even sell your excess electric generating capacity to your local utility.

To learn of many of these products, how to get them, And how to earn those rebates, goto: **http://www.ecoenergyanswers.com**

One Final Warning

I often read about coming alternative fuels and systems that will be available in thirty or even fifty years.

We do not have that kind of time to wait!

Do the math. At the rate petroleum prices are rising (long term) and at the rate our purchases of fuel from the oil-rich nations are growing, our economy could be destroyed long before we have converted to new energy systems. The 2008 economic downturn could be the first indication that it is already happening. If we do not develop our own economic lifeblood to replace petroleum fuels within a much shorter period our economy will have bled to death. I believe ten years for almost complete replacement of petroleum fuels is close to the maximum time we have left. Even that period will be fraught with painful economic problems.

Low-cost energy is essential for us to maintain our vibrant, growing economy. Oil has gone from ten dollars a barrel to one hundred and thirty dollars in the last ten years, and dropped to fifty dollars a barrel in 2008. The biggest fluctuations (read that as instability) have occurred in the last three years. Should the long-range trend continue, the price of gasoline could pass fifteen dollars a gallon by the end of 2013. By that time our economy will be losing as much as $4 trillion annually to the oil-rich nations. Unfortunately, our economy will have collapsed (and our oil companies will be out of business) before that point is reached and that is only four years away. It is truly amazing that this tremendous drain on our economy has not already precipitated a major depression. It is certain that such a collapse is imminent and will probably come before we can complete the switch to alternate energy systems. In light of current economic conditions (2008) the question arises, could such a collapse already be underway at least in part because of the billions of dollars for oil already drained from our economy? The only way to reduce the severity of the damage to our economy is to make the changeover quickly. The sooner we do so, the less damage to our nation's and even the world's economy will occur.

Section IV

Politics Rears its Ugly Head

A Sad State of Affairs

The reality of politics and political ideologies means that many politicians and bureaucrats, who know virtually nothing about energy, energy systems, and the economics of energy, will be making many of the decisions on what systems we use, the vehicles we drive, and how we create and pay for the new infrastructure. Political benefit and not practicality will probably be the leading driving force. Because of this it is quite unlikely the best and most efficient systems will prevail. For detailed information explaining why this unfortunate information is so true, read John Stossel's book, *Give Me a Break*.

Note for political ideologues: I respectfully suggest that if yours is the closed mind of the political ideologue, and comments that disagree with your beliefs and agenda offend you, even after reading John Stossel's in depth information, it might be prudent for you to skip the next few chapters and go directly to the Appendix on page 187. I am a believer in American style free-enterprise capitalism and a rational, logical approach to solving problems. I also believe profits are wonderfully beneficial for most, if not all, Americans. I base the following pages on that belief. This could color the opinions of readers who do not share my views. Most of the content of the book outside of Section IV is nonpolitical and should be of interest to all points of view.

Though politics is a major player, the purpose of this book is not political. It is written for the layman in everyday terms to provide information and encouragement for the public. Yet it has enough technical information to interest those able to make things happen—the thinkers, doers, and movers of our nation. The numerous energy systems described include those used for several hundred years, those just discovered and in their infancy, and even those long known but grossly underutilized. Many of these systems will fall into disuse or be maintained merely for historical or sentimental usage. Others will move to the fore and become important factors in the energy economy of the future. It is my hope that activists and rank-and-file, from across the entire political spectrum, can shelve their powerful prejudices, and take a pragmatic view of the very real and useable systems proposed in this book.

Success in improving our energy systems or creating new ones would greatly enhance the economic, environmental, social, and human factors in our nation and even the rest of the world. Our position in the world and the respect of people everywhere would surely follow.

Some Personal Experiences

I have a few friends and even some among my family who are so imprisoned by far out ideology that they will not read, listen to, consider, or even discuss anything, not within the confines of their *box* of accepted values or actions. I have even been ordered by a few of them not to send or try to discuss any of my opinions that differ from their positions on anything. The rest of them merely ignore me, or avoid any discussion. It is interesting to note, that I have been described as an extreme liberal by some of my conservative friends, and as a right-wing fanatic by some of my more liberal friends and family. These otherwise fairly normal folks are true fundamentalists of both the extreme left and extreme right, some political, some religious. They readily express hatred toward those who disagree with their *holy* positions in any way. I truly feel sorry for the bitterness they must feel to be so adamantly against things they hardly understand. Most of their opinions are repetitions of mantras or slogans promulgated by leaders of groups or mass movements they belong to, or follow.

The author knows much has changed since he wrote the major portion of this book. The mortgage and financial collapse, the major national election changes, the recession that the anti business, high-tax rhetoric of the new administration has caused to deepen, the government takeover of most of the financial and auto industries, have each contributed to drastic change. These together with the promise of much higher government spending, debt and taxes all have significant negative effects on business, investment, and jobs. Investment in new technologies, so desperately needed to solve our energy problems, has been a leading victim of these negative influences. Motivation for creative innovation has been badly damaged and may not recover for an extremely long time, if ever. Those creative entrepreneurs, who invest in new technologies, new industries, new products and economic progress, are certainly looking for action in places more hospitable than the new United States.

I now look for the ideas expressed in these pages to be developed mostly in the Far East in places like China, India, and Korea or elsewhere like Ireland, and possibly even Russia. Even South and Central America looks to be poised to move from stagnant socialist economies to the excitement and promise of entrepreneurial capitalism. It would not surprise me to see many of our young entrepreneurs looking to those nations moving steadily from socialism to capitalism for their futures, while leaving a United States heading in the opposite, dull gray and unproductive direction. Because of these negative economic factors, principal organizers and investors have put on hold or cancelled many of the American projects mentioned in these pages Entrepreneurs are simply not going to risk until they know they have a chance to make money. The threat of massive new taxes on the *wealthy* along with the hostility toward business that is freely expressed by liberal Democrats, is a powerful detriment to entrpreneurial risk..

It is my opinion that the thousands of independent American golden geese have been severely wounded. The current political atmosphere is certainly not conducive to their recovery. Just look at what is happening to the financial condition of the three states of New York, California, and Michigan. Many businesses in those states have failed or moved to more business friendly climates in other states or nations. The effects on the economies of those states are devastating and yet they continue spending and still speak of increasing taxes. This costs them more businesses, more jobs and more lost revenue. It's a vicious cycle they can't seem to understand or will not. It is my hope that the American people realize what is going on and call a halt to this foolishness.

The recent sound defeat of ballot initiatives in California gives some hope this may now be happening. Also, the surprising results of the election in Massachusetts show that some people have realized what a destructive force liberal policies are. Unfortunately, this may come too late to reverse the damage already done, especially the just passed Obama-care bill, which is far more about power and control than health care.

America is a large, friendly dog in a very small room. Every time it wags its tail; it knocks over a chair.

Arnold Toynbee

Changes in the World Economy

The recent mortgage banking meltdown is just one example of how inscrutable each of these factors of change can be. It also shows just how much damage political manipulation of economic conditions can create. Apparently, a group of self-serving political appointees replaced sound banking practices with politically expedient actions and *cooked the books* of two huge government mortgage companies to hide the results. Couple that with the application of government pressures on mortgage bankers to make bad loans, and finally the bubble burst. This was the straw that broke the back of the much abused mortgage system. Suddenly, billions of dollars of equity disappeared, almost overnight. As a result, the government's two mortgage companies collapsed bringing down with them several major mortgage banks. The ripple effect triggered the collapse of several other giant financial institutions before expanding into the general economy.

At the same time, months of $4+ per gallon gasoline added another negative effect on the economy. This and the constant harping on the *terrible economy* by politicians and the media destroyed public confidence and dealt the final blow to two of America's big three auto makers. Combined with the mortgage debacle, this brought about an immediate and dramatic downturn to the economy. The drastic drop in demand drove the price of oil below fifty dollars a barrel for the first time since 2004. This has resulted in the pump price of gasoline dropping below $2 per gallon in many places. Demand for many things ground to a halt as mortgage foreclosures skyrocketed, people stopped buying, jobs disappeared, and unemployment rose abruptly. With American retail sales dropping catastrophically, those nations that have been supplying America with consumer goods are suddenly in an economic free fall. The resulting drop in demand for petroleum products, in those nations, further precipitated the slump in petroleum prices. This led to a substantial drop in orders for expensive goodies for the oil sheiks. The snowball effect on the world economy has stock markets plummeting all over the world while investment in business expansion virtually ceased. None of this bodes well for the economic future.

Few people realize that the mortgage meltdown created the largest transfer of wealth out of the hands of individual Americans in history, mostly middle-income Americans. Combined with the constantly repeated threats of the administration against business profits with increased taxation, this has caused small and medium sized businesses to cancel all plans for expansion and new employment. At the same time, many are downsizing and laying off employees in preparation for the growing economic depression. Indeed it is already a *tax* depression directly created by the policies of the Obama administration and liberals in the federal government.

How does this affect the move toward alternative fuels? It's obvious that with gasoline hovering around $2 per gallon, the economic pressure to switch to alternative fuels has dropped substantially. Alternative fuels costing $2 and more per gallon that looked attractive when gas was $4 per gallon, will encourage no investment now. It could be that the low price of petroleum products will call a halt to the development of alternative fuels. With the only incentive being the questionable but powerful global warming movement, it is highly unlikely investors and entrepreneurs will look favorably at alternative fuels, at least for the present. For the same reason, drilling for oil in places like the Bakken formation in North Dakota and Montana, will not look nearly as attractive for investment as it did when oil prices were higher. Already these facts have caused T. Boone Pickens to shelve plans for an enormous wind-farm in Texas. His liquid natural gas project will continue as planned since the price of natural gas is dropping just like petroleum.

On the other hand, electric vehicles, batteries, and electric power generation will remain attractive if only for the low cost per mile they will offer, even lower than petroleum fuels at the new lower prices. Add to this, the long life and low maintenance cost associated with electric vehicles and they look more like the best promise for the future. With new battery technology being the heart of an electric vehicle system, and new types of electric power generation supplying the blood, these systems look to this writer to be by far the best bet for the future. Shrinkage of dollars available for investment in new products and systems occasioned by the economic downturn will certainly slow the development of needed products for an EV/PHEV-based economy.

Government Involvement

Government involvement can have a significant positive or negative effect on any development. Government meddling was at least complicit in generating the mortgage meltdown, and the resulting banking failures. Now they are getting into the auto business via the bailout route. This could throw up a sizeable barrier to new vehicle development by artificially sustaining some financially troubled auto companies with taxpayer dollars. Be reminded that the big three are only a part of the auto industry in this country. Chapter eleven bankruptcy would certainly have enabled them to regroup, retain their assets, and return to competitive viability. It has done so in many other industries like the airlines. Of course, one could argue that government interference and regulations brought about lots of unplanned expenses of retooling, and because of that, perhaps the government could share some responsibility for their poor financial state. This could justify at least some of the bailout money they have managed to obtain. Government uses money to bring about the changes they want that may or may not be good for the nation as a whole. All of this is free of any of the economic pressures of profitability.

Let's examine a scenario based on the success of EVs, PHEVs, and massive new electric generating capacity. Suppose that by 2018 we have nearly doubled our electric generating capacity, and that nearly half of our vehicles are running on electric energy. At today's prices for electricity, private electric vehicles cost about 4¢ per mile and commercial trucks about 16¢ per mile. The associated drop in demand for petroleum fuels would result in about the same cost. The maintenance advantage of the electric vehicles over those powered by fuel would be substantial, a major factor in corporate decisions about fleet purchases. Some of the side effects would be closing about half of our refineries and the related loss of jobs throughout the oil industry. Those workers would have to be retrained for new kinds of jobs. A significant part of the automotive repair and parts industry would also disappear even as some switch over to maintenance of electric vehicles. Serious disruptions would occur among suppliers and others that support petroleum fuel systems. While the energy industries, and the repair and maintenance companies, would certainly be American, the vehicle manufacture could go overseas. They are certain to go overseas unless they were persuaded economically to develop here

Most of this scenario will only be possible with a complete and permanent reversal of the government's hate campaign against business, capitalism, job providers, and the *wealthy*. These are the economic entities that make things happen, create new products and industries, create new jobs and drive a successful economy. That is certainly unlikely to happen as long as liberal Democrats are in power.

One could go through each of a dozen factors with any of the situations described and find differences and problems we would face because of the possibilities each could present. One quick look will provide a sense of the complexity of these problems.

In addition, any rapid response to these problem situations presupposes a major commitment by many people and some active cooperation between groups and organizations currently at each other's throats in angry verbal and political combat. Continuation of existing politically determined policies and activities will merely get us more of what happened that brought on this crisis. That includes higher fuel prices, even higher food prices, and no new oil, oil alternatives, or energy sources—an economic catastrophe. Trillions of dollars will continue to bleed from our economy and into the coffers of the oil-rich nations, many of whom work for our destruction. What we need is a creative new direction for our nation, mostly, a reversal of the destructive new policies that put us where we are. A large part of the world is presently going in the wrong direction. We should be leading in a new and dynamically positive direction and not following that old, worn path to socialist tyranny.

Politicians and Political Forces

What is that wrong direction and what is this new and better direction? What I am about to say and propose could turn off many people. In most instances, this includes those who have the closed minds of the extreme left, the extreme right, or any other controlling ideology. It will do the same to those who are emotionally bound to any ideology, the nonthinkers and the blind followers, whose minds are so addled by passion they have few moments of rationality. These are the destroyers of the world, the angry permanent children, who learn that destroying things of value gets them the attention and power they so desire without having to think, plan, or work very hard. From their ranks come the likes of Hitler, Stalin, Bin Laden, gang leaders, political hate mongers, and the angry mobs of the world. There is no reasoning with these individuals—no negotiation—no cooperation—no significant communication—certainly no peace in any sense of the word. They understand only power and destruction, anything that uses fear to control, sometimes even to the point of murder of innocents. To quote an American phrase, "It's their way or the highway."

Although these people are the ones who could most benefit from the ideas in this book, few such strong ideologues will ever read it because it does not fit precisely into acceptability in their closed minds.

The author is hopeful those with open minds, those who recognize truth and reality, will read and use this information about the extreme complexity of the interrelated issues and possible solutions outlined throughout this book. There will certainly be a chorus of practical answers along with many impractical ones. Let us hope we pick from among the most practical and effective. The following is about one nation that cast off ideology, adopted a realistic political approach as described herein, and took a fresh, new approach. Just look at what happened.

The Shining Example of Ireland

Ireland is a prime example of what can happen when government frees businesses and entrepreneurs from oppressive controls and taxes. At the same time, those public and media persons who held visceral hatred for business, profits, and *the wealthy* abandoned their unreasoning hatred and became not necessarily pro business, but at least not so angrily anti-business.

After years of wallowing in poverty, in a country where government controls and high taxes on business stifled progress and discouraged investment of both time and money, the Irish took a dramatic new course. Government interference and controls of business were largely abolished. Complex reporting, that bogged down management, was mostly thrown out. Corporate taxes, once among the highest in the world, received a significant change in the late 1970s. Things changed when the government included capital gains in corporate profits, and reduced the tax rate to 32%. Then in the 1990s they reduced the corporate tax rate to 12.5%. Government changed from being the enemy of business to being a strong supporter. The results speak for themselves. This so stimulated the Irish economy, that between 1981 and 2001 corporate tax revenues tripled. As a result, Ireland became the poster nation for lower corporate taxes. Ireland is now one of the most vibrant economies in Europe. Business is booming like never before. There are now many high paying jobs, and investment capital is flowing freely into a nation that once couldn't coax any investors. In the last twenty years, more than 1,000 foreign companies have moved to or opened operations in Ireland. Local firms have also flourished and expanded with worldwide impact. Employment has grown so much that Ireland now imports thousands of workers just to keep their industries running. All of this success is primarily because of the new, positive attitude of the public, the media, and government of Ireland to the development of business. This radical new public posture has brought on the availability of world class support services including banking, trade finance, transport systems and advanced telecommunications.

During the world economic downturn of 2008-9, Ireland has fared badly because of their dependence on foreign economies. Much of their unemployment is in construction and foreign commerce where multinational companies have reduced their spending on facilities.

Irish political scientist and historian, Tom Garvin, is the author of several books on Irish political history. "I have to make a mental effort to remember the Dublin of the 1950s, which was in many ways a Third World city," Garvin reports. "Horses, no motorcars, children in bare feet, dirt everywhere, people living in slums, no television, no bathrooms—an impoverished European country that didn't seem to be going anywhere." When a dear friend visited Dublin in the 1970's, she found little improvement from Garvin's picture of the 1950s.

When she and I visited Dublin together in October of 2007, the city was almost unbelievably different: vibrant, hopeful, optimistic, enthusiastic, and almost ecstatic. Expensive clothes and new vehicles were everywhere. The local car dealer had buyers waiting sixty days to take delivery of their new BMWs. People filled the streets and literally overflowed the pubs and upscale stores that lined them. This amazing economic outcome resulted from government working with business rather than against it, and removing oppressive tax burdens rather than imposing them. What also helped was a pro-business attitude of people and even the media, rather than the class hatred and anti-business attitude we see so prevalent in our own country today.

Doing something right for a change: Ireland is a good example of what can happen when warring factions lower their weapons, tone down their rhetoric and do just that. Fifty years ago it was virtually a third-world nation with much internal strife. Today its rapidly expanding economy gives testimony to the wisdom of pro business government attitudes especially in an impoverished nation. Ireland did what I suggested as my fondest wish for America. The Irish example led me to believe so strongly that government can favor business and the worthwhile jobs businesses provide while at the same time dealing successfully and reasonably with the wealthy, and helping the poor learn how to help themselves. Government can also limit the danger of monopoly and the *robber barons* that gave business such a bad name in the late eighteenth and early nineteenth century. (It looks like those American *robber barons* have now moved from industry into the halls of Congress and Washington's nightmare bureaucracy.)

Mostly, the Irish changed their attitudes—a lot. Someone once said, "Attitude is everything." Another old saying is, "You can catch more flies with honey than with vinegar." Simply stated, Ireland did just that. Intelligent actions based in these types of positive concepts had a lot to do with Ireland's meteoric rise. Generally speaking, their people changed their attitude about business and industry—big time. In their government, in their media, and among their people, they changed from anti business to being predominately pro business. They lowered business taxes and removed burdensome controls and restrictions, especially those that didn't work or were a source of political payoffs. As a result, businesses prospered and grew, investment money poured in, much of it from the US, and their economy took off like a skyrocket.

We should learn and profit from their example. What we need now is a lot more honey and a lot less vinegar used among politicians, government, the media, business, and even the public. The rapidly expanding economies of not only Ireland, but China, India, and many other smaller nations give testimony to the power of a friendly attitude toward business, toward profits, toward investors, toward capitalism, and toward workers.

The Political Challenges We Face

John Stossel, consumer advocate and 20/20 reporter, explains that much of what we hear—and what the media say—are myths. In his book, *Myths, Lies, and Downright Stupidity, Get Out the Shovel Why Everything You Know Is Wrong*, Stossel points out how politicians and activists use anecdotal evidence to prove the truth of concepts that are factually untrue and often downright harmful. Here is one of the nearly two-hundred common myths debunked in his book. This one is in the chapter about business.

MYTH: Government must make rules to protect us from business.

TRUTH: Competition protects us if government gets out of the way.

It took me a long time to learn that regulations can't protect consumers better than open competition. After all, I worked in newsrooms where *consumer victimization* was a religion and government its messiah. But after fifteen years of watching government regulators make problems worse, I came to understand that we didn't need a battalion of bureaucrats and parasitic lawyers policing business. The competition in the market does that by itself. Word gets out. Angry customers complain to their family and friends; consumer reporters like me blow the whistle on inferior products and shoddy services. Companies with shady reputations lose customers. In a free society, cheaters don't thrive for long. *(At least not in business.)*

Once I learned more about economics, I saw how foolish I had been. Government uses force to achieve its ends. If you choose not to do what government dictates, men with guns can put you in jail. *(And clever lawyers will remove gobs of money from you to get you out.)* Business, by contrast, cannot use force, no matter how powerful they are. So all business transactions are voluntary—no trade is made unless both parties think they benefit. In 1776, economist Adam Smith brilliantly realized that the businessman's self-centered motivation gets strangers to cooperate in producing a multitude of good things: "He intends only his own gain, and he is in this, as in many other cases, led by an invisible hand to promote an end, which was no part of his intention.

Few of us appreciate the power of that invisible hand. I don't give my pencil a second thought, and yet I could spend years trying to produce one without turning out anything as good as the worst pencil available.

He goes on quoting an essay *I, Pencil* by Leonard Read of the Foundation for Economic Education. The essay describes the people equipment and organizations actively involved in gathering the components of the pencil from all over the world. This includes many items: the machines and workers who combine those components, the machines and workers who make those machines, the trucks and ships that carry the raw materials and finished product around the globe, and the systems that distribute the pencils to the end users. It's an amazing and enlightening bit of prose. I highly recommend reading Stossel's books. They will provide an instructive background for how to accomplish much of what the information contained in these pages promises for America. I also recommend viewing one of his talks about freedom and free enterprise that can be viewed at:

http://video.google.com/videoplay?docid=1876894381231272307

Sadly, politicians will have much to say about what we do and how we address the growing problems involving energy. These problems include many that have serious economic effects now and will in the future determine the health of our economy. These politicians have little understanding of energy problems and solutions, or the invention, development and manufacture of fuels and energy systems. As a result, they are poor judges of what might and might not work. The reason for the political emphasis in this section is to discuss the problems facing any new technology and business entity, that must run the political gauntlet.

Politics is a totally emotional game with virtually no rational component. Many politicians are totally engrossed in ideology and are without any practical ideas or effort that does not support their ideology. They are far more likely to propose and vote for those things that promote their agenda or benefit their backers and constituents rather than practical or creative proposals that address the problems. Entrepreneurs and their investors will be the ones who will solve these problems if government just gets out of the way.

Unfortunately, politics and politicians will have a great deal to say about where we go with energy, which systems get *approved* and funded, and which are discarded. This is unfortunate because many crucial decisions will be made by people who have no understanding of the real actions, values, or cautions required by any energy system. Their decisions will almost certainly be made by those who provide them—the *governmentalists* (my apolitical term with obvious meaning)—with the most power and/or money in response to their almost pure emotional appeal. Politics has so invaded many of our institutions (education, unions, entertainment, the media, and many professions) and is so controlling, that little can be accomplished without the money to buy political clout. Because of efforts of the left, government has grown from 4% of our economy just before Lyndon Johnson's administration to more than 40% as this is written. Unfortunately for Americans, Alexander Tyler's and Henning Webb Prentis's comments and predictions (page 134) have certainly proven accurate for America.

No one can accuse any politician of providing an honest, forthright, nonemotional proposal on anything. In fact, even to question some of their pronouncements—to ask for answers to rational questions and concerns—is to invite public ridicule at the hands of politicians and celebrities including the main-stream media. The public's growing worship and adoration of celebrity dooms us to succumb to the charms of charlatans who promise the moon while lining their pockets and empowering themselves. The ancient worship of *royalty* is not dead. It is merely substituting a new set of *royals* so enamored of the media celebrity worshipers.

It is my considered opinion and experience, that most Americans can rise above these unreasoning passions if given the opportunity to open their minds, to think and reason. I am a sincere believer in the basic goodness of most people, Americans in particular. I have seen powerful evidence of this in the outpouring of compassion accompanied by free giving of physical aid during times of catastrophe virtually anywhere in the world. This has often been twisted by those who envy and thus despise America. Recently, after a major natural disaster, one news service reported the US government had given far less in disaster relief funds than the governments of a number of smaller nations. The truth, that was never mentioned, was that the private, nongovernment gifts from Americans dwarfed the private giving of the entire rest of the world. When added to the government's gifts the total was more than the next three or four nations total giving combined. That is just one example of how a hostile news service can twist the news and skirt the truth to serve their particular agenda.

Free, independent Americans are the ones who should find interest in this book, and for whom it is written. These are the open-minded thinkers and doers, workers and organizers, creators and builders, who made this country the best, most free and independent nation on earth by their energy and hard work. I urge those who belong to this shrinking majority of real Americans read with an open and objective mind. I'm sure many will find statements that run counter to long held personal beliefs, no matter where their political loyalties lie. When this happens, please read the entire section as objectively as possible and consider the logic of the arguments presented. The understanding gained will be surprising.

Many new and revolutionary materials, systems, and combinations of these are now available that could solve our energy crisis in just a few years, a decade at most. We must investigate these and bring the best through development into production and use. This will only happen to those that gain attention and favor in the eyes of the influential and then the public. The best will fall by the wayside if they do not gain this favor, no matter how excellent. Unfortunately, politicians, the media and entertainment people, who are mostly ignorant of the complexities and nuances of energy problems and systems, could have a lot to do with the choices made that will impact our immediate future considerably and our long-range future drastically. Entrepreneurs will also have a big part, but only if their efforts are rewarded and they are not shackled by emotionally driven government regulations. The next chapter, starting on page 139, deals with these realities in our nation as I see them.

It is my fondest hope and wish that, within our nation, we can find cooperation and respect between groups that now usually see each other as enemies, or sources of political power rather than fellow Americans trying to do something positive by building for a mutually advantageous future. The groups I speak of include academia and their excellent researchers, private enterprises from individuals to the largest corporations, governments and government agencies, local and national, and all of the entertainment world including the media, Hollywood, New York, and the world of sports. I realize this is a big order, particularly since the political wars have ratcheted up to fever pitch and emotions run high. The human energy and dollar expense, consumed by these growing conflicts incited by angry rhetoric, are enormous and terribly wasteful. Emotions and resentment for past real or imagined injustices are powerful and deeply held. Nevertheless, peaceful cooperation and a little understanding and tolerance along with some give and take can work wonders.

Doing things exactly the wrong way. Growing rancor of political campaigning is one example of the power of inciting hatred to sway voters. The preponderance of personal attacks over substantive proposals shows how much easier it is to tear down a political opponent, than to build one's own stature by making serious proposals. Negative campaigning is easy and especially effective in the age of the *sound bite*. Like any mob action, it is easy for idiots, the uninformed, and the easily led to join and support because no thinking is required. Any dolt can do that with little training. In political campaigns, it is pure emotion that mob leaders use to drive voters. Serious proposals, even solutions for all-important problems, rarely get the media play and public attention that hate-filled rhetoric aimed against those same proposals receives.

For similar reasons, it takes far less skill or organization to demolish a home or even the World Trade Center than it does to conceive, design, and build the same thing. Conflict is easy. Cooperation and creative building are far more demanding, require careful consideration, dedication, creative effort and hard work. They are infinitely more rewarding. This is what we sorely need right now.

Often attributed to Lincoln in error are these words penned by William J. H. Boetcker, in 1916.

You cannot strengthen the weak by weakening the strong.

You cannot help small men by tearing down big men.

You cannot help the poor by destroying the rich.

You cannot lift the wage earner by pulling down the wage payer.

You cannot keep out of trouble by spending more than your income.

You cannot further the brotherhood of man by inciting class hatreds.

You cannot establish security on borrowed money.

You cannot build character and courage by taking away a man's initiative and independence.

You cannot help men permanently by doing for them what they could and should do for themselves.

Yet are those not precisely the short lived, instant gratifications politicians and media personalities regularly wield against those they oppose for any reason?

A prediction of where we seem to be headed may have come from far back in history, when the 13 colonies were still part of England. This quote is often attributed to a Scottish Historian, Alexander Tytler or Tyler. The true origin of the quote is obscure and might have originated in the early 20th century from an unknown politician or writer. Nevertheless, this does not detract from its accuracy.

One version of this quote on why democracies always fail is:

"A Democracy cannot exist as a permanent form of government. It can only last until the citizens discover they can vote themselves largesse out of the public treasury. After that, the majority always votes for the candidate promising the most benefits from the public treasury with the result that the Democracy always collapses over a loose fiscal policy, to be followed by a dictatorship, and then a monarchy."

A version of the second part of the misquote, often attributed to Arnold Toynbee is:

"The release of initiative and enterprise made possible by self-government ultimately generates disintegrating forces from within. Again and again, after freedom brings opportunity and some degree of plenty, the competent become selfish, luxury-loving and complacent; the incompetent and unfortunate grow envious and covetous; and all three groups turn aside from the hard road of freedom to worship the golden calf of economic security. The historical cycle seems to be: from bondage to spiritual faith; from spiritual faith to courage; from courage to liberty; from liberty to abundance; from abundance to selfishness; from selfishness to apathy; from apathy to dependency; and from dependency back to bondage once more."

But the person who appears to be the actual author of the second part is Henning Webb Prentis, Jr., President of the Armstrong Cork Company. In a speech entitled, *Industrial Management in a Republic*, delivered in the grand ballroom of the Waldorf Astoria at New York during the 250th meeting of the National Conference Board on March 18, 1943, and recorded on page 22 of *Industrial Management in a Republic*, Prentis had this to say:

"Paradoxically enough, the release of initiative and enterprise made possible by popular self-government ultimately generates disintegrating forces from within. Again and again after freedom has brought opportunity and some degree of plenty, the competent become selfish, luxury-loving and complacent, the incompetent and the unfortunate grow envious and covetous, and all three groups turn aside from the hard road of freedom to worship the Golden Calf of economic security. The historical cycle seems to be: From bondage to spiritual faith; from spiritual faith to courage; from courage to liberty; from liberty to abundance; from abundance to selfishness; from selfishness to apathy; from apathy to dependency; and from dependency back to bondage once more.

"At the stage between apathy and dependency, men always turn in fear to economic and political panaceas. New conditions, it is claimed, require new remedies. Under such circumstances, the competent citizen is certainly not a fool if he insists upon using the compass of history when forced to sail uncharted seas. Usually so-called new remedies are not new at all. Compulsory planned economy, for example, was tried by the Chinese some three millenniums ago, and by the Romans in the early centuries of the Christian era. It was applied in Germany, Italy and Russia long before the present war broke out. Yet it is being seriously advocated today as a solution of our economic problems in the United States. Its proponents confidently assert that government can successfully plan and control all major business activity in the nation, and still not interfere with our political freedom and our hard-won civil and religious liberties. The lessons of history all point in exactly the reverse direction."

These are the real malignancies we must overcome if we are to solve the rapidly growing problems, facing not just the US, but the entire world. The ultimate collapse of America will be death and destruction of unprecedented magnitude unless we reverse course soon.

The Realities of the Gulf Oil Disaster

As this is written, it is more than four months since the explosion that resulted in the sinking of the deepwater oil platform. This loosed a torrent of raw oil and gas from the mile deep floor of the Gulf off the shore of Louisiana. The resulting string of stupid blunders by BP and the companies running the oil rig combined with the bureaucratic ineptitude of our competing government agencies created a nightmare of accusations, claims, and counter claims by many parties. In addition, long delays held up most actions for a variety of ridiculous reasons, including jurisdictional disputes between government agencies, domestic and foreign companies and unions. Example, the sand berms to keep the oil from gaining access to extensive Louisiana marshes: After a rather long delay, the federal government finally granted permission and Louisiana started pumping sand to build the berms to keep the oil from reaching the marshes. Shortly after the pumping started, environmentalists convinced the government the pumping might endanger the Chandeleurs, the sand islands that offer some protection to the Louisiana coast over time. The government stopped the berm building until a study could be conducted to determine the environmental effect of moving the sand. Who knows how many months that study would take. It would probably be little more than a series of ongoing arguments between conflicting views and come to no viable conclusion.. In the meantime, the oil damage to the vast wetlands will probably cause a thousand times more environmental damage than the worst the sand pumping would do.

This is but a single example of the many skirmishes occurring over what to do and how to do it. It's a classic case of *too many chiefs and too few Indians* typical of the ridiculously complex, conflicting agencies of our obese government. Each agency and often each head of this or that trying to make themselves more important. All this goes on with total disregard for the costs paid by taxpayers. To add to the confusion, many if not most members of both private and public organizations have little understanding of their own rules. Another example, also from the oil spill, is the various interpretations of the Jones Act of the 1920s by various individuals.

Witness words from one private source: "The Jones Act only applies within three miles of shore. Therefore, foreign skimmers, along with American skimmers, are already at work beyond three miles. The Deepwater Horizon spill is occurring 50 miles from shore, and the vast majority of oil is beyond 3 miles."

From another source: "I believe you will find the Jones act applies to the *territorial waters* of the United States. In the 1920s, when the Jones act was written into law, those territorial waters extended 3 miles from shore. Since 1982, the United Nations law of the sea defines *territorial waters* as 'a belt of coastal waters extending twelve nautical miles from mean shore line.' These waters are considered sovereign territory of the state owning the land. In response to several nations (including the US) that arbitrarily declared the extension of their *territorial waters* 100 to 200 nautical miles from shore, the UN defined an *Exclusive Economic Zone* extending 200 nautical miles from the defined shoreline. In this zone, 'a coastal nation has control of all economic resources including, fishing, mining, oil exploration, and responsibility for any pollution of or by these resources.' The Jones act is considered by many as covering this economic zone."

This plethora of confusing authorities, both written and personal, that are negatively affecting the efforts to fix the oil spill, is typical of both government and corporate bureaucracies. It's everyone trying for a share of the credit when things go well, and everyone pointing fingers of blame at others when things go wrong. All of this is fueled by greedy lawyers eager to sue anyone with deep pockets on any side of any controversy, and media journalist personalities more interested in sensationalism and their own agenda than facts. Frequently, we ignore viable solutions to the real problems or lose them in the melee and uproar of conflicting accusations. It is certainly easier to blame someone else for a problem than to find a solution.

Add to this confusing mix a president with virtually no experience at running or managing anything other than a political campaign. He is a master of platitudes and euphemisms, an orator who is quite obviously without a clue as to how to manage or direct any real situation. I only mention this because of the tremendous power the federal government now wields which must be considered in any energy decision about solutions to anything. It seems a group of grammar school kids and schoolyard bullies are now running things in Washington.

A tempest in a teapot? Suddenly the hand wringing prophets of doom and gloom have grown silent about the Gulf oil disaster. With the shrimp fleet now operating, the fishing fleet now free to work, and scientists unable to find even traces of oil or chemical dispersants in Gulf waters, one wonders, what happened to *the worst environmental disaster in history?* Certainly searchers can find a few trouble spots where mangroves have partially died and coastal swamps remain a bit soiled by old oil, but even these are rapidly disappearing. We can thank mother nature and her oil eating microbes for most of this. In retrospect, it may be that the most disastrous economic effect to the region was the loss of tourism and vacation business. This was certainly brought about mostly by weeks of media predictions of long term oil contamination in the Gulf, vast beaches soiled with ugly oil, and waters too contaminated to swim or even travel in a boat. Add the many *hate business, hate oil companies, hate America* activists and politicians, and it becomes a formula for economic destruction. *News* people seem incapable of objective reporting. Sensationalized disaster and tragedy reporting is their stock in trade. Objectivity and pragmatism is not. If the first reports make a disaster ten times worse than it is, the next reporter will make it a hundred times worse, and the next a thousand times. It's just the nature of the beast. Yes, it was a major disaster. There were people killed, lives and businesses were disrupted, the Gulf environment took a hit, and this writer would like to know what actually happened at the Horizon Deepwater drilling platform. However, as virtually always, the disaster was infinitely smaller and the damage immeasurably less than the news media would have us believe.

Exaggerated condemnation and name calling is counterproductive unless the desired result is conflict and destruction. It's about time members of groups that constantly denigrate and condemn others out of class or economic envy, or those with whom they have a jurisdictional dispute, began to honor and respect the achievements and accomplishments of those others. I see it as essential that we recognize the realities of our situation and the real reasons we are where we are. There has been enough of this destructive and debilitating blame game, and all of its political distortions and emotional, hate-filled activities. We Americans engage each other in terrible inner political warfare, while our enemies stand on the sidelines urging on the various sides, and gleefully watching our self destruction.

This book includes many links to blogs and web sites. The information available on the Internet is far too voluminous to be included. It is information that anyone can access using a computer. Numerous web sites and news reports disappear soon after being posted for many reasons including obvious inaccuracies, editorial revisions, and/or real changes that make them invalid. A single powerful caveat exists about Internet information. Many charlatans lurk well hidden behind attractive facades on the Internet. Even a cursory examination will reveal there are magic, energy out of nothing, and secret smoke and mirror articles readily available. Many of these are driven by hidden agendas. Every searcher must be certain that information comes from a reliable source, and truly is from that source. All information must be checked and confirmed from other reliable sources. Even those sources can be in error for many reasons. In addition, the often carefully hidden agenda of these sources and their record of reliability must be considered.

The following section deals with political, jurisdictional, corporate, and communications problem in greater detail. This is the background of where we are, why we are there, and what we must do to go forward. It has little to do directly with energy and fuels, but everything to do with what we must overcome in order to counter this growing menace before it destroys us.

It is my personal conclusion that government and corporate bureaucracies along with political and media activists are among the most powerful forces working against real solutions to the serious problems facing humanity. Rather than use those problems as a focus for knowledgeable people to find creative solutions, these negative groups use them as bludgeons with which to punish and if possible, destroy those who do not support their agendas. The vast majority of ordinary citizens of the world are swayed this way and that by intense and usually hate-filled rhetoric of persuasive individuals in politics and the media. It is no wonder they get angry because of frustration at leaders who quite obviously are feathering their own nests at the expense of the common folk while blaming all the economic woes they have created on those, like the *wealthy* who have achieved more. Of course, they exempt themselves from being among the *wealthy* no matter how much power or wealth they possess.

Why any Changes in the Fuel/Energy Industry Will be so Difficult and Costly to Accomplish

Why we are where we are with regards to petroleum.

As much as we would like to have one, there is no quick fix to our dependence on increasingly expensive petroleum. It will take years of expensive research and development to find and develop the best fuel replacements for gasoline and diesel fuel in enough quantity and at a competitive price. In the interim, existing fuels will have to suffice. Consider the following:

1. **The world demand for petroleum fuels is growing rapidly with no end in sight.** China leads the rapidly developing world in increasing that demand and surpassed the US in oil consumption in 2008. Over the long haul, this drives the market price of petroleum higher worldwide with a temporary drop in demand caused by the recession. World production of crude has begun to slow as existing oil fields are becoming depleted and production is slowing. New oil fields are harder and harder to find, and many known fields are too expensive to tap, especially at today's lower prices. No American oil company can do a thing about that.

   a. The three largest oil fields under production in the world are past their prime and production is declining and getting more costly. These three include, Saudi Arabia, Kuwait and Mexico. All things being equal, this alone would cause the price of crude to rise.

   b. Control of nearly 94% of the world's crude oil production is in the hands of foreign oil companies. Many of them are state-owned monopolies. Most of these state owned oil companies have used their oil as a cash cow for the government, and have not set aside cash reserves to search for new oil to replace their dwindling resources.

      (1) Pemex, the Mexican state owned oil company needed to explore for new oil to replace the rapidly diminishing Cantarel oil field. Since they had squandered all of their oil profits on handouts and civil projects, they didn't have the estimated 18 billion dollars to commit for the needed exploration.

When the government proposed an investment that would have included private participation in Pemex, a partial privatization, leftist activists raised demonstrations and riots against it, so the plan was scrapped. This clearly shows how short-sighted, ignorant, and self-serving leftists are, even to their own detriment. Faced with the slow deterioration of their only consistent cash source, the government of Mexico has done nothing. They can only obtain the eighteen billion required by increased taxation. This will further damage an already declining economy. Maybe Mexico should lower taxes like the Irish did and thus increase revenue.

2. America has a virtual sea of oil within its borders and around its shores. Thanks to what I believe to be misdirected effort to influence elected officials by some overzealous environmentalists, the most accessible of our known oil fields are off limits to American oil companies. Others are so expensive to drill into that they will never be a significant source. The Chinese are now tapping some of those offshore oil fields. They do not have the relatively accident-free record of our drillers. They are not controlled by the high level of safety and environmental regulations as are our drillers.

   a. Drilling horizontally, the Chinese can access oil fields just a few miles from our coast. American oil companies could be pumping these same fields using American labor, generating American profits and paying taxes to the American governments and maybe lowering our pump price of fuel. **So much for the wisdom of our politicians.**

   b. A blowout of one of their wells or a major leak in one of their pumping stations could seriously damage our Gulf Coast. We would have no control over such a tragedy as we would if it were our oil companies doing the drilling. **Now the Gulf oil disaster says, so much for federal controls on deepwater drilling.**

   c. Environmental activists are trying to protect the herds of Caribou by preventing drilling in ANWR based on the theory our activities would affect them negatively. These Caribouo are actually helped by the heat of the very same kind of oil pipelines already in place. The Caribou huddle near these heated lines in extremely frigid weather for survival. It is interesting to note the Caribou herd has increased dramatically since the oil fields and pipeline were built. **So much for the claim of harm to the caribou herd.**

   d. American oil companies have become expert at drilling undersea without incident or spill. The many drilling rigs damaged by hurricane Katrina did not spill a drop of oil into the Gulf. OK, maybe that is a bit of an exaggeration, but no blowouts or significant spills occurred, even in the face of those devastating hurricanes.

That is until the disastrous blowout of the Horizon Deepwater well in April of 2010.

e. The largest known oilfield in the world is under most of North Dakota, Montana, and Saskatchewan in Canada. This field, called the Bakken formation in the Williston basin, has been known since the 1950s. Drilling for and recovery of this oil is quite expensive. Drilling is estimated to cost 15 to 30 dollars a barrel. Therefore, using the field was just not cost effective when crude was between 10 and 20 dollars per barrel. Now, with the price going through the roof, this more costly oil is becoming more attractive. Those who know estimate it will take the investment of billions of dollars to bring this field into full production and even that will take several years. Realistically, we may be able to extract only a tiny portion of the *sweet* crude oil contained in this field because getting it is so difficult and expensive. Out of the 400 billion barrels some estimate is contained in the formation, a mere one to one and a half billion appear to be extractable economically by current technology. I only explain about this to show just how difficult it is becoming to find new, viable sources for crude oil. For more information on this potential source of crude, go to Internet site:

http://tonto.eia.doe.gov/ftproot/features/ngshock.pdf

f. So, for the time being and to get us over the rising costs of petroleum and petroleum products until we have viable alternatives, I join the chorus saying, **"Drill here! Drill now! Save money!"** That means environmentally safe drilling on the North Slope including ANWR, the continental shelves, the Gulf of Mexico and anywhere else we can find domestic sources of oil. If we do not, our economy could soon be destroyed by fuel prices going sky high. Billions of dollars will then leave our economy and build the wealth of despots, who hate us and vow our destruction. One prominent North Dakota state official put it this way, "We are working hand-in-hand with members of the oil industry to turn the potential of the Bakken field into reality. We find them knowledgeable, cooperative and hard working. They are providing good jobs and economic growth for our state while demonstrating genuine concerns for the environment. They are indeed friendly partners in our state's economic growth rather than the evil *Big Oil* as painted by others." That certainly is a far more realistic policy than the animosity toward business and *Big Oil* expressed by the media and Washington elite.

3. **Business and in particular, Big Oil, is getting a bad rap from some politicians and the media, as usual.** During a recent Congressional investigation into high fuel prices, several Senators demonized oil executives who make a lot of money. When

asked to explain his multimillion dollar compensation, one executive pointed out that business executives competed for jobs just like other workers. His salary was in line with other executive salaries with similar responsibilities and was not nearly as high as some sports figures or entertainers. Also, his company's profits, 11%, were not as big a percentage of sales as Microsoft, 34.24%, Apple, 18%, or Pfizer, 17%, for example. I wonder why the Senate never investigates or condemns the millions paid to sports figures, rock stars, or other entertainers. Why not accuse them of *gouging* the public?

a.   Think about it! The price of fuel at the pump is but a single example of how real world events have a major effect on the wealth of our nation as a whole. These events impact our living standards, our environment, and our physical and emotional health. Business is getting a bad rap from the other groups because business is a tempting target for attacks motivated by class envy. One Democrat Senator made the revealing comment, "Democrats love employees, but hate employers." The often repeated phrase, "Everybody hates the boss," is the result of constantly inflamed employee/employer battles. This helps take public attention away from the mischief happening within the political arenas and the ill-informed public lap it up eagerly. Make no mistake, we find cheats and scoundrels in all groups including businesses, governments, academia, labor unions, and others. Some members of these same groups are also afflicted with nepotism, favoritism, immorality, discrimination, and self service. None are immune to these human frailties. In my opinion, the financial competition in American business will soon weed out those with negative effects on profitability or their companies will go broke. Add the intense scrutiny of government and the media to the financial operation and business must spend a great deal of effort just to keep their record clean. Many succeed and for very long periods of time. Read and listen to consumer/taxpayer advocate John Stossel if you want to learn the political and economic realities few Americans understand. See section V starting on page 189 for links to his books and Internet video presentations.

b.   I wonder if the fortunes amassed by so many in Congress or the other branches of government could stand up to the kind of scrutiny imposed on business executives. An examination of the finances of politician and government officials, that moved from modest economic condition to extreme wealth after a few years in office, could be very enlightening. I'll wager much more of this kind of mischief, including illegal activities, are going on in political and other public endeavors than in business. That is because commerce itself holds business to account by the need for profit and financial integrity just to survive. Politicians and other groups have no such limits. Politicians buy votes with government money and no one objects or even seems to care. The *Robber Barons* of the late eighteenth and early nineteenth century have indeed left the business world and now sit in the halls of congress and populate much of our federal bureaucracy.

c.   So who and what is really to blame for high fuel prices? It's not Big Oil that prevents drilling in ANWR, along our coasts, and virtually anywhere else in the sea of oil in and around our country. It's not Big Oil that threatens massive new taxes on fuel. It's not Big Oil that is raising petroleum consumption in China, India and elsewhere around the globe. If government really wanted to do something positive, legislators or executives could rescind that part of the forty to seventy cents per gallon tax that does not go into repair and maintenance of our roads and highways. That is a whole lot more than the five or ten cents per gallon oil companies take as profit. Add the constant harangue of class envy against Big Oil, in the media and from politicians, and it is no wonder the public has little concept of whom the real culprits are. I am no big fan of oil companies, but they are definitely getting a bad rap that should be equally charged against the media, politicians and those opportunistic environmentalists. These self-serving activists are preventing us from drilling for the abundant oil in and around our nation and doing so just to serve their own political ambition and agenda. Still, domestic oil production would only be a short range, temporary solution until alternate fuels could be developed.

d.   An effort on everyone's part to lower taxes and abolish or at least diminish this contrived animosity. That could go a long way to help solve this dangerous energy crisis. If this happens, it might begin to draw business that has moved overseas back home. The benefits, even to a less intrusive government, can be substantial. Ireland's tax revenue soared because government removed the ridiculous, make work regulations on business while at the same time taking a much smaller portion of the pie. As a result, the pie became so very much bigger. Which is the better choice, high taxes and a small pie or small taxes and a big pie? In other words, would 10% of $1,000 or 2% of $10,000 yield the better return? Investment capital, the lifeblood of any private or for profit enterprise, always flows to where the biggest returns can be made. There is nothing complex or emotional about that. It's axiomatic. Time again in this country the proven effect of tax policy on tax revenue is the same. In every case, lowering the capital gains tax results in booming business and increasing tax revenues. In contrast, raising these taxes results in business slow downs and lower tax revenue. The Kennedy, Reagan, and Bush tax cuts all resulted in increases in federal tax revenue.

If you were an investor, would you move your money to the U.S. with a promised 40% to 50% corporate tax and a 39% capital gains tax, or Ireland with a 12.5% corporate tax that includes capital gains? It's no wonder the stock market is plummeting. Investment money is fleeing our shores for better returns elsewhere. The prediction attributed to Professor Tyler was right on the money. We are presently doing exactly what he foretold.

e. In the face of these facts and in spite what it means for our nation, the new administration and Congress promise a doubling of the capital gains tax and the removal of the Bush tax cuts. To add to our nation's problems they are proposing additional new taxes to pay for expensive federal programs and bureaucracy. All this will do is add to our economic woes. These same people also prevent us from constructing needed new refineries, or from building new nuclear and coal-fired power plants. As a result, we are handing increasing billions of dollars to nations whose leaders hate us and vow our destruction. How anyone can propose such actions is beyond my understanding. Apparently, they want so badly to punish successful American people for being successful they are willing to sacrifice job growth, tax revenue, and the economic health of the nation to do so. These proponents of economic disaster either are stupid as a log or have some sinister but hidden purpose. None of their proposals make rational sense. It is the use of political power at its worse and promotion of class envy as a motivator. The result will be that everyone suffers and could ultimately lead to the economic destruction of our nation.

Our enemies could hardly plan a better strategy!

Quotes from Arnold Toynbee:

Civilization is a movement and not a condition, a voyage and not a harbor.

Civilizations die from suicide, not by murder.

Civilizations in decline are consistently characterized by a tendency toward standardization and uniformity.

Some Personal Observations

I would like to make certain my beliefs, thoughts, and ideas as well as their background, are known. Knowing the mind set, beliefs and understandings of the author will make the content of these pages more clear and understandable. Make no mistake. I am an independent believer in America, American excellence, the American Constitution as it was written, and American capitalism. I do not try to second guess those great men who founded this nation and created its Constitution. Their words and intentions are quite clear to anyone who is familiar with our Constitution, or has read their various papers. I am a conservative in that I know what the word *conserve* means, how to do it and its limitations. I understand that real freedom is priceless. It requires constant diligence to protect free men from demagogues who would take away their freedom in the name of security. Those who want to be taken care of from cradle to grave can find just such a place in many totalitarian nations, or even in our prison system. Slavery is another name for that kind of security.

While reading this section, keep in mind that much of this has little to do directly with any solution to our growing energy crisis. It does, however, have a great deal to do with how we tackle the problems, who tackles the problems, which systems gain favor and use, what we do about fuel and transportation during any transition, and how quickly we solve these problems. Most of all, government will never solve these problems. People—independent, entrepreneurial people—will, if only government stays out of the way.

This section has to deal with that premise. It examines many people, organizations and systems that will influence how we attack and hopefully solve the current energy crisis. This is done by explanation and by providing examples. This background has little to do directly with the subject matter of the book. It does have much to do with why it was written, why politics will have a great deal to do with the choices made, and why it will be so difficult to solve the energy crisis. Any solution must run the gamut of political prejudices, business hierarchies, public clamor, and the obvious biases of a media far more interested in celebrity, scandal, sensation, and their own political agenda than fact. Those able to gain or buy favor with the afore mentioned will gain the most attention. The squeaky wheel is still the one that gets the grease.

It is my opinion that most humans are inherently decent creatures whose nature is to care for and respect at least some of their fellow humans. This ranks high among the reasons humans have been so successful in survival and reproduction. Even among the groups of evil people I use in the horrible examples are doubtless some kind, thoughtful, well intentioned if naive individuals, usually well hidden. Sadly, others are unkind, devious and have evil intentions. We have all met both kinds, but mostly we deal with the infinite variations between those extremes.

I also want to make clear my feelings and opinions about political positions, parties, and the like. I support ideas and proposals that make sense to me and are congruous with my technical training and experience as well as my moral beliefs. Capitalism, even with all its highly criticized evils, is far more conducive to the well being of the largest percentage of the people than socialism or communism at their best. It is certainly far better than any kind of totalitarianism. I also believe the old saying, "That government governs best that governs least."

There are many evils the left attributes to capitalism. I believe those evils are actually faults of human nature and are present far more in the hierarchies of socialism than in the U.S. under our form of capitalism. The actual dispersal of power is the reason. In socialist nations, the power to control and govern including the means of production and distribution is always in the hands of or under the control of a single group. With socialism there is little or no competition, no individual free-enterprise ownership or involvement, and no incentive to excel, improve, or be efficient.

Actually, socialism is a modern version of the European feudalism of the middle ages, complete with lords and ladies—the politicians in power—and serfs—the common folk. Socialism, marxism, communism, feudalism, and liberalism are merely names for totalitarian political systems that are virtually identical in practice and result. Socialist governments, except those that are very small, always grow their bureaucracy to the point at which the productivity of the nation drops to low levels and the living standard of the common people suffers greatly. Sooner or later their despotic leaders recognize the power of the media to mold public opinion. They then take control of the media, silence dissent, and tell the people only what they choose to let them know. North Korea is an example of this in the extreme while Venezuela is rapidly going the same way. Are we next?

Contrast the organization in socialist nations where a single group calls all the shots with the many thousands of stockholder-owned companies we have in the U.S. Each company has its own independent board of directors, executives, and stockholders who vote and so control the actions and the futures of their companies. I can guarantee those stockholders have more to say about the operation of the company whose stock they own than the voters in any nation have to say about the operation of their government. Capitalism is the ultimate form of public ownership of the means of production and organized services. Socialism is the form of bureaucratic control of the means of production and organized services by a single, small group of politicians—the ultimate monopoly. The large number of independent and self actuating organizations in any American style capitalist nation spreads control over a wide range of individuals and organizations. There are evils, scoundrels, and lazy nonproducers associated with our form of capitalism, but certainly far less of these than in our government. They pale in significance compared with those found in socialist monopolies.

Most socialist nations of the modern era, evolved from one of two very different types of states by several very different means. The first is from nations previously under totalitarian control. Usually an uprising overthrows a hated king, dictator, or corrupt government and takes control of the country. Those in power during the uprising may have high ideals and intentions, but eventually corrupt, power hungry, and usually ruthless groups take over the government. This usually ends with a new totalitarian state with a single dictator. The old Soviet Union, China, North Korea, Cuba, and Venezuela are all perfect examples. Iran is an example of the Islamic fascist version of socialist dictatorship which acts much like a socialist government and controls virtually every Islamic nation. These governments are examples of what happens when any group or entity has a monopoly of power and is accountable only to itself. Corruption at all levels of government becomes rampant. This results in an economic collapse when the cost of maintaining the closed economy outruns the ability of the people to produce. It can also be the result when the source of natural wealth of raw materials dries up. The Soviet Union, Cuba and North Korea have already passed through this phase. The Islamic dictatorships will face the same collapse when their oil runs out, or they lose oil revenue for any reason. This is manifestly true because they are not developing any other sources of wealth or wealth creation.

The second type is where a free democratic nation chooses to move to the left because voters elect those politicians who promise the most from the public treasury. As the government expands, promising more and more largess to most of the people, the economy eventually collapses because of fiscal stupidity. There has never been an exception to this. It happens when the wealth of the people and their ability to produce is taken away by steadily increasing taxes levied by a leftist government. Mexico, Greece, and a number of South American countries are in this process as this is written, Several other European nations are well into the same path, with Spain the next one likely to drop over the fiscal precipice. Should the US continue on the path taken by the leftist Democrats elected in 2008, economic disaster will soon follow. The left achieves equality for all when everyone except the politically powerful becomes a member of the dependent poor.

I recently visited the former East Germany where significant evidence of the destructive power of socialism still exists. Agriculture there is but one powerful example of why human nature makes socialism a destructive, negative force, and free enterprise capitalism by far the best economic system for the most people. The East German government took all the land away from the farmers and turned them into collective farms where the former farmers were employed by minions of bureaucrats and their quota systems. Good old, *From each according to his ability, to each according to his need.* took control, and all incentive for personal thrift, industry, creativity, and hard work evaporated. Agricultural production in the entire area never reached a third of what it was prewar. Waste, sloth, lack of creativity, and laziness became the rule because—why bother? The thousands of individuals who were killed trying to escape East Germany provides a tragic example of what a socialist state must do to keep its citizens from escaping to freedom. Since the fall of the East German state, and the reunification with West Germany, the collective farms were broken up with most being returned to their former owners. The rest were sold to new owners. The result? Agricultural production is now many times that of the old collective farms. Industrial production is following a similar path. This demonstrates conclusively the vast superiority of capitalism over socialism, which is, in reality, just medieval feudalism under a new name.

China, India and even Russia are three examples of socialist nations that have discovered the power, enthusiasm, and energy that come from even a modicum of capitalism. Each has responded differently. China has partially embraced capitalism with much enthusiasm and is continuing to privatize more and more of its considerable production capabilities. Their current leader, a scientist and engineer, reportedly announced, "Profits are good! Business is good! Free-enterprise is good! Capitalism is good!" That is an unprecedented reversal of form and verges on the unbelievable. Tom Friedman describes these amazing facts in his book, *The World Is Flat*.

This all started years ago when Chinese officials realized that farmers cultivating their own tiny fields were out-producing the collective farms by a large margin. They were doing so with more and higher quality produce, and in their spare time. These tiny capitalist enterprises were far more productive than the state owned farms. This is almost universally true. Governments have almost never been able to do anything productive as well as private enterprise, not as economically, not as efficiently, not as profitably, and government spends tax dollars while private business pays taxes.

In these now rapidly expanding economies, ownership has given their citizenry the incentive to work hard and care for the fruits of their labor. Having *a piece of the action* has been and will always be a powerful motivator to excel and produce high quality goods and services in the most efficient manner. No government entity can possibly compete with men who are free to use their own effort to provide for themselves and their families. They never have and never will. The burgeoning Chinese economy is evidence of the power of free-enterprise capitalism, even in a totalitarian state. How this will play out politically in the future remains to be seen. Once the free-enterprise genie is out of the bottle, it will be difficult if not impossible for the communist government to maintain their absolute control and stuff those freedoms back into that bottle. The heady experience of economic freedom and the resulting life style once tasted will be hard to squelch peacefully.

Our federal legislature will be another major force involved in the selection of America's new energy policy and the industries it influences. Special interests and their lobbyists, powerful individuals with huge fortunes, self-serving politicians, pork-barrel projects and earmarks, each have far too much power in Washington. They will have a major influence on where we go with energy. Members of the Washington in-crowd have developed using public funds to buy votes for incumbents into an art. It is not necessary to mention the countless blatant examples here. The media have come to ignore these evils because they have become so common place.

Pork barrel projects and earmarks are the means politician's use to buy votes while sometimes lining their pockets at the expense of the public. They use your federal tax money to buy local votes from their constituents. They spend these tax dollars to provide profits for favored businesses. In addition, they trickle down to some local jobs. Of course, the job's part is all they ever mention. They never talk about the profits made by any of the following: contractors that are frequently relatives, businesses they own, the increases in value of property they own, consulting fees, or jobs with lobbyists for relatives. Those politicians who have been exposed are just the tip of the iceberg. Why are there so many nouveau riche in Congress? If that is not because of well disguised political payoffs, I'm the next President of the United States.

While liberal Democrats are by far the most abusive, members of Congress from neither party hold a monopoly on these practices. They frequently brag to their constituents about their actions. (Vote buying?) They are seldom if ever called to account for their actions because all of them do it. Makes one wonder about the bigger fish that are never *caught with their hand in the cookie jar*, does it not? For whatever reason, public figures seem able to get away with far more evil without complaint than private individuals in business. Unfortunately, these self-serving politicians will impact the outcome of any effort at new energy.

The Obama administration has raised this practice to a new level with the *stimulus package* which some have correctly renamed the *porkulus package*. They used virtually all of this money to reward party faithful and secure votes and support for administration policies. They even used the TARP money and several *bailout* actions to the same end. The bailout of the auto industry was purely a bailout of the UAW who now own and run GM (Government Motors) and Chrysler. Even the closing of auto dealerships by the administration was purely political. They closed most successful dealerships, owned by

Republican supporters, while maintaining those that have provided Democrat support. Among the dealerships maintained, were many that were failing financially. They shut down virtually all of the 100 most successful GM and Chrysler dealerships in the nation by arbitrary executive order without recourse. Does that seem rational to any thinking person?

Those wonderful new powers that be at GM chose to trash two divisions, Pontiac and Saturn, rather than sell them to willing buyers for a whole lot of cash. Could it be they would rather waste all that taxpayer's money than let all that value to end up in private, capitalist hands?

Several significant things happened as a result. Ford has realized a dramatic increase in sales as have several foreign auto makers. GM and Chrysler sales have plummeted as buyers confidence in those company's products virtually disappeared. Why? Would you invest in an important product being manufactured and supported by the same people who are running AMTRAK, the Postal Service, Medicare, Social Security, etc., etc.? As long as there are competitive products and services available from reliable, accountable organizations who must make a profit to survive and prosper? I certainly would not.

There are many examples of the epidemic of abuses involving the use of public funds by our elected officials. These abuses have reached ominous proportions. Another example of these unstoppable bureaucratic juggernauts is the latest farm bill. To subsidize farmers, to the extent this bill does, is ridiculous if not immoral. Especially at a time when the price and profits from grain used in the expanding manufacture of alternate fuels are rising precipitously. Supposed to help the family farm, these subsidies go mostly to gigantic corporate farms who fund and control the farm lobby. I wonder what route the money takes to reach the coffers of the politicians who approved that bill.

Money for politicians will be a major factor in determining just what fuels we use as alternates and the industries and systems required to supply and use those fuels. It will also be a major factor in any changeover to alternative forms of energy. I only mention these things because they are a reality any organization or person will face in any efforts to develop and market alternative fuels or energy systems.

That being said, I provide these words of warning. The media and many politicians are doing their best to make American oil companies, Exxon in particular, extreme villains for raising the price of gasoline at the pump. That is specifically because such commentary serves their agenda and removes the heat from their actions no matter how far it strays from the truth. Those who promote this idea either are completely ignorant of business and the oil business in particular or know they are lying just to shift the blame from themselves.

With the new giant oil spill in the Gulf, British Petroleum has become the new whipping boy of the left. Quite obviously they are far more interested in condemning BP than in getting the oil leakage stopped, or in finding out what actually caused the explosion that precipitated the crisis. This tragedy, no matter how big or small an environmental disaster it eventually becomes, will be used to leverage massive oil company controls and taxes while painting those *evil* oil companies as terrible villains.

The cause of this disaster has not yet been thoroughly explained, nor has the media said much about a possible cause. One can easily conjure up many scenarios involving various kinds of terrorism. This media silence, as well as the absence of any explanation about the SWAT teams the administration sent to other oil rigs in the gulf, makes me extremely suspicious about the actual cause of the disaster. Could they simply be hiding the real culprits? Could environmental or Islamic terrorists have caused the explosion to produce precisely the situation we now face? That oil rig was a sitting duck out in the gulf, a tempting target for those eager to cause problems for America and America's supply of petroleum. Is it not very strange that any accidental explosion could have damaged that platform in such a way as to sink it in such a short time? With all the safeguards in place to prevent fuel explosions on every oil drilling platform, such an accident is very unlikely. It is far more likely that a specific high explosive was placed so as to sink the platform quickly and bury the evidence under a mile of ocean. It would be very easy for a team of scuba demolition experts to accomplish this from a nearby fishing trawler and get far away before the charge went off. In the absence of detailed information from the media about the cause of the blowout, countless scenarios scripted by the many enemies of America's oil industry can be imagined.

The stake of all American oil companies combined in the world petroleum business is a mere 6%. If they shut down all of the wells they control tomorrow and went out of the oil exploration and drilling business, oil producers from the Middle East, Venezuela, and Russia could pick up the slack in a heartbeat. Saudi Arabian oil companies alone could deliver that amount and more. These are not publicly-owned companies. They are state-owned monopolies directly under control of their political leaders—socialism in action. Sometimes the profits go directly into the pockets of the rulers. This is the case in virtually every middle eastern oil-producing nation as well as Nigeria, Venezuela, and even Russia. The largest oil company in the world is now PetroChina, a state owned company. These oil companies have infinitely more control over fuel prices than do American companies, yet our media and politicians continue to demonize our own oil companies while ignoring those who do control the world's oil. I wonder why? This is not to say or even infer that some of our own oil companies do not play dirty for their own benefit.

What are the long-range goals of these new world oil companies and the leaders of the oil-rich nations? Whatever those goals are, one of the most troubling results is the transfer of trillions of dollars from Western economies to the personal holdings or control of despotic leaders. Even the dimmest light bulb among the American public could understand the menace of the growing mountain of debt now held by the Oil sheiks and China and their banks. Many are now suggesting that American oil companies are to blame. They can't get at the oil sheiks or Chinese so they attack a convenient domestic entity because it suits their political agenda. Some are even suggesting the answer is for government to take over the oil companies. That would be one more step in the destruction of economic freedom in the United States.

Any government takeover of a business or industry would be a monumental disaster for that business or industry and for the nation. Once on that slippery slope, Marxists in our government would soon turn our marvelous, free, productive, diverse, and colorful nation into a dull, colorless imitation of the old Soviet union.

The fat little socialist dictator in Venezuela did just that in his nation. He used class hatred to inflame his people and bring down the ones who built the Venezuelan oil industry. He now personally controls this huge source of cash, after taking over the infrastructure of several American oil companies that had invested hundreds of millions in the Venezuelan oil industry. Chavez told them to grant his government controlling interest in their installations or get out without compensation. Several oil companies have just walked away, choosing to do so rather than commit to helping Chavez. Like many on the left, Chavez is buying public support with public money and by selling gasoline for mere pennies per gallon to Venezuelans. It seems to work for him. Recently, Chavez shut down the only remaining voice of dissent in Venezuela. Since then he has installed a system of individuals spying on their neighbors. Should these neighborhood spies not report on their neighbors they are imprisoned. It looks like Venezuela is heading in the same direction as Cuba did fifty years ago. This is a good example of real socialism in action. It sounds a lot like Hitler's Germany, or Stalin's USSR to me.

Please keep all of that in mind while reading on about several ominous situations that are affecting our economy right now. These are situations that will seriously impact where we as a nation go with energy and fuels, and how the choices we make will change our economy for better or worse. The size of the problem and the positive and negative economic changes our nation could face are enormous, even catastrophic. One wrong choice, one error in reaction, could destroy us economically in a short time.

We live in a society and culture of bicameral differences, even where there are many more than two positions. We are clearly divided into dichotomous groups: male and female, young and old, child and adult, Democrat and Republican, even liberal and conservative. These differences divide us by race, culture, ethnicity, religion, politics, language and any number of physical, social, financial, ethical, professional, and emotional factors. Those who would control or influence us have developed sophisticated ways to use these differences as divisive factors to set us in conflict with each other.

Virtually all questions and problems morph into simplistic pro and con arguments, yes versus no, right versus wrong, black versus white, etc., when the actuality is far more complex involving many shades of gray. It has been asked, why is it we cultivate our few differences to divide us rather than use the vast number of things we have in common to unite us? The answer is quite simple. There are those who know how to divide and conquer, and use that as a means to their own purposes.

Divide and conquer has been the heart of modern politics as it has been in warfare from prehistoric times. Divisive hate speech, accusations and counter accusations, these are the heart of presidential campaigns and most political contests. Political campaigns provide many emotional diatribes, few if any substantive proposals or commentary, and almost no truth or facts. It is no-holds-barred emotional combat.

Mass movements drive much of this. They are emotionally charged and determined by slogans, heroes, villains, and countless blind followers playing follow-the-follower. Even though started with the best of intent, mass movements can change and be driven by hatred. This almost always leads to major human disasters, even when they succeed in their avowed purpose. Too frequently the movement takes on a life of its own, far from the original intent no matter how lofty that intent was. Many examples of this exist along with many reasons why it is so. Eric Hoffer described the emotional followers of so many passionately driven mass movements many years ago in the following words.

"Passionate hatred can give meaning and purpose to an empty life. Thus people haunted by the purposelessness of their lives try to find a new content not only by dedicating themselves to a holy cause but also by nursing a fanatical grievance. A mass movement offers them unlimited opportunities for both."

Freedom and open, objective access to education and information are the enemies of those who command these mindless, purposeless forces. That is because those who are truly free know that freedom demands responsibility for one's own actions. Freedom must be taught. Even though it is among the most basic of human animal instincts, real freedom is frequently overwhelmed by conformity and peer pressure at an early age. Therefore it must be taught at those early ages by parents and teachers who understand and value it. Another Eric Hoffer quote illustrates my meaning.

"People unfit for freedom who cannot do much with it are hungry for power. The desire for freedom is an attribute of a *have* type of self. It says: leave me alone and I shall grow, learn, and realize my capacities. The desire for power is basically an attribute of a *have not* type of self."

The enemies of freedom—those who hunger for power—are those who would impose their will, their values, their mores, their political system, their political correctness, and even their religion including atheism: on everyone under their control. In many organizations and nations, those in power make this imposition with force or even threat of death.

No matter how benevolent they are at the start, movements that follow these patterns ultimately turn violent and evil. The Christian church of the middle ages, the industrialists of the eighteenth century, the labor unions of the early twentieth century, the Nazis, even organized crime, the Mafia, and modern city gangs, are examples. Islamic fanatics have lead a devilishly cruel and deadly movement since coming on the scene in the seventh century. Certainly not all involved were evil men. It's just that the infatuation of force usually brings *the meanest son-of-a-bitch in the valley* to the fore as the leader, Such leaders know dead opponents never cause trouble. These leaders are not far removed from the *king* of a monkey troop who fights his way to that position over the mangled and dead bodies of those who oppose him. We humans, on the instinctive (emotional) level, are not so far removed from our simian cousins.

The Communist revolution in Russia is a classic example. Started by a large group of idealist intellectuals, the Bolshevik revolution toppled a weak and intransigent royal government. Once they had defeated that government, the idealists and intellectuals gave way to the vicious Stalin dictatorship. In the process, virtually all of these idealists and intellectuals were brutally murdered or thrown in Siberian prisons. Leon Trotsky was even tracked down to Mexico and murdered by Stalin's assassins. *The meanest son-of-a-bitch in the valley* had indeed risen to be the monkey king of the Soviets.

John Emerich Edward Dalberg Acton, first Baron Acton (1834 1902) made the famous statement,

"Power tends to corrupt, and absolute power corrupts absolutely. Great men are almost always bad men."

The last part of his remark is rarely mentioned but it is certainly true. I wonder, why is this so? Our American Constitution was written by truly great men who were only bad men in the eyes of the British. These brave Americans knew first hand the evil power of oppressive, controlling government. That document was crafted carefully for the sole purpose of protecting free men from their government, even though it was a republic. Far from perfect, that document seems now to be under attack by those who would *modernize* it or change its meaning outside the methods outlined in that document. The desire of these favored few individuals, is to gain control over people, and force them to submit to their will. Like the Taliban, they seek to destroy any object or person that does not support their *sacred* way. Free Americans must be vigilant to keep these *favored few* from using the courts to reinterpret the Constitution to achieve their self-serving purpose.

Politics is strictly an emotional game. As a consequence, the most important factor for any American candidate to have now is celebrity and the support of the media. The entertainment-driven media thrives on celebrity. As a result, many Americans have come to value celebrity and image above all else. The public seems to value celebrity above honesty, character, ability, education, experience, even morality. These all take a back seat to celebrity in virtually every election. This is most unfortunate because celebrity and leadership ability rarely go together. Just because an individual can do the things that create their celebrity status like, make a stirring speech, do a magnificent job of acting, have a beautiful face or body, sing like an angel, or throw touchdown passes: these attributes do not make them wise or enable them to be a capable leader. Witness the large numbers of celebrities, usually from the entertainment world, who take up causes and speak out with apparent authority about things of which they know little or nothing.

Liberal-socialist establishments, organizations, and leaders here and throughout the world, constantly seek control of communication and entertainment. This control enables them to manage words as propaganda—informative, influential, persuasive, inciting, and condemning, but always serving their agenda. Their propaganda is mostly emotional or instinctual—passions, hatred, laughs, and tears—with little or no factual content. Frequent use of words like *more, better, change*, or *fair* move some people emotionally, while having absolutely no real meaning as used.

They also seek influential power over education at virtually all levels. They want to mold the thoughts, concepts and beliefs of children and young people to serve their agendas. The decimation of our education system including poorly equipped graduates, gives testimony to the damage leftist policies have wrought on our once unexcelled system of education.

Other examples of the power of indoctrination of children and young people via mostly leftist educational institutions can be seen in the news or found in history books. This is especially true as history books are rewritten to be politically correct. Of course, *politically correct* should more properly described as liberally or even tyrannically correct and LC should replace PC. Important stuff is now left out while PC stuff is added. Chris Columbus now landed on an island of peace-loving natives and enslaved them to get gold. Never mind that the Caribs were aggressive cannibals who had wiped out and eaten many of the other tribes in the area and proceeded to wipe out the small garrison ol' Chris left behind (in righteous wrath, I'm sure).

Most people my age learned about the Hitler youth movement, as it was happening. We heard how young Germans were taught they were superior and that Jews and others were inferior. The German puppet masters then blamed all of the woes of the German people on these people and created the holocaust. Certainly that is an over simplification of the reality, but it did happen almost that way. But of course, Hitler's Germans were pikers at murder and mayhem compared with the hordes of Islam.

Islam, the brutality of Muslim fundamentalism

Muslim beheadings and terrorism are nothing new. These emotionally driven fanatics have been committing those atrocities on any who would not convert to their faith since they first came on the scene in the seventh century. Madrassas in the Muslim world have taught millions of young people only from the Koran avoiding any other source. This has created countless angry followers who have been taught only hatred toward anyone who does not adhere to their beliefs. They learn from nothing else, from no other source. Only a privileged few, mostly the wealthy and powerful, receive any other education at all. Is it any wonder that Islam has had a bloody history of conflict with and invasion of their neighbors?

Muslim invasion of India—Few people know that while the Muslims invaded Persia in 634, they invaded Sindh in India in 638, a gap of just four years. While Persia succumbed in seventeen years by 651, Muslims took seven hundred years to overrun India (today Sindh is a part of Muslim Pakistan that was carved out of Hindu India in 1947). Even after that, they could not rule India in peace. Over the centuries, mobs of Muslims frequently attacked non Muslims and especially, Hindus, hacking them to death with knives. Victims included men, women, children, and the aged. They are continuing the same thing today as witness the 2008 attacks in Mumbai. The only difference is now they attack with bombs and bullets (and civilian aircraft). When they acquire atomic weapons, they will not hesitate to use them to the same purpose.

Historians have written about *Islam's bloody borders* over many years. Charles Krauthammer said in an article titled, *The Bloody Borders of Islam*, published in the *Tampa Tribune* on Dec 6, 2002. "From Nigeria to Sudan to Pakistan to Indonesia to the Philippines, some of the worst, most hate-driven violence in the world today is perpetrated by Muslims and in the name of Islam." The complete article can be viewed at:

http://www.freerepublic.com/focus/news/838321/posts

The Hitler youth movement and Islamic Madrassas are just two examples of how indoctrination of the people can work to serve the purposes of the few puppet masters who control masses of blind followers. This ultimately results in the game of *follow the follower*, human lemmings willing to do virtually anything including sacrificing their miserable lives to serve the *cause* of their masters. These masters will do whatever it takes to keep their followers poor and ignorant and control them. They do this by offering up enemies that are made responsible for their misery. This includes encouragement to have many children to feed their armies of blind followers. Many of these blind followers will gladly sacrifice their lives for their leader and against their supposed enemies.

I only describe these horrors because, for the most part, these people are the ones who control the world's oil and whose despotic leaders are receiving the mountains of dollars being spent on petroleum by the rest of the world. They are the ones who will eventually rule the world unless we quickly develop the alternate fuels described in this book. Make no mistake. Their long-range plan is for your grandchildren to be taught only from the Koran under threat of the scimitar. That is if they manages to live through the takeover. The growing cost of petroleum plays right into their plans as do the efforts of those who are preventing us from drilling for our own oil. The only way to stop them is to remove the bans on drilling in our domestic oilfields, while feverishly working to develop alternative energy sources and fuels, and fast. This is an infinitely greater and more impending menace than supposed global warming could ever be, even at its worse. If the current rate of the expansion of Islamic population continues, within twenty to forty years, most European nations will be Islamic and under Sharia law.

The Growth of Socialist/communist Power in America

Currently, members of the new administration and Congress are announcing their plans to impose many more taxes and more government controls limiting our freedoms and controlling the actions of individuals. At this point in time, the camel has his head in the tent and is poised to move in bodily. Up until the new administration took control in Washington recently, our freedoms were not disappearing suddenly, but in tiny, virtually unnoticeable incursions like the camel sneaking into the tent. Since then, the rate has accelerated noticeably and shows no signs of slowing. Many successful industries (like drugs and oil) have been a target for hate speech and condemnation by liberal Democrats with their typical class envy tactics. While both industries show billions in profits, their actual profits expressed in percentage of sales or investment are quite modest compared with other industries, around 11% for oil (below average) and 17% for pharmaceuticals (above average). Why aren't those liberals in the media screaming at the *obscene* profits of Microsoft projected to be $19.34 billion or 34.24% profit for 2008?

Enemies of capitalism seem always to use dollars when they want to condemn major corporations and percentages otherwise. Does the public even have a grasp of how much a billion dollars is, or that billions in total profits may represent only an average return on investment? Do they understand that those profits are usually reinvested to provide new opportunities for American business and new and better jobs for Americans? Every dollar taken from those profits as taxes is a dollar that will not add to business development or payrolls. The people in Ireland learned and embraced that understanding. This turned their nation into the most dynamic economy in Europe. The amazing thing is that just a few decades past they were virtually a third world nation.

Some politicians are bludgeoning the oil industry with class hatred, saying that oil companies should be taxed to pay the federal fuel tax of eighteen cents per gallon to reduce the cost to consumers. Seeing that the oil companies net only about a nickel, or at most a dime a gallon on fuel and that governments combined tax is from forty to seventy cents per gallon, their claim and suggestions are preposterous. They carefully hide the government's take of eight to fifteen times that profit to keep the public's focus off of the taxes. That is why it is actually government, not the oil companies, that gouges motorists.

If the government puts the oil companies out of business, or takes them over, as one mindless congressional representative suggests, who's going to pay to replace the billions in taxes provided to the government by those oil companies? Last year, those taxes represented $28.5 billion paid by Exxon alone. This amount is almost exactly the amount the lower income half of American wage earners paid in income taxes in 2007. Think of it, one oil company, Exxon, is paying the same amount in income taxes as the entire lower half of American wage earners. Still, oil company profits have been around 11% of sales and 8% return on investment in recent years, a modest percentage comparable to the average of other businesses.

Democrat lawmakers are calling for new *windfall profits* taxes similar to those signed into federal law in 1980 by President Jimmy Carter. These would tax the profits of major oil companies at a rate of 50 percent. If so, why limit it to oil companies? Why not tax all corporations at 50% or maybe 90%? That's one sure way to destroy America's strength and productivity. Does anyone remember the unemployment, mounting inflation, and interest rates of 22% that Jimmy Carter's financial wisdom gained for the country during his presidency? Quit the class warfare and glue sniffing, people, and get real.

Who Actually Pays All Those Taxes?

The truth of the matter is that few members of the public understand the realities of any taxes. Every business, large or small—every professional—every taxable entity or organization—every worker, all appear to pay taxes to many governments. Actually, corporations do not pay taxes. Businesses do not pay taxes. Professionals in their professions do not pay taxes. Organizations including those who are not-for-profit do not pay taxes. **If that is so, who is it that actually pays these taxes?** The truth is, all of these *taxpayers* merely collect those taxes from customers or clients in the final price to the consumer of their products and services and then pass them on to the government. They collect taxes exactly like the much more obvious sales tax. Even the so-called company-paid portion of FICA taxes add to the cost of an employee and so are actually taken from the employee.

What happens if lawmakers enact windfall profits tax at 50% of net profits? It's quite simple. Every dollar increase in taxes will be met with a slightly larger total increase in fuel prices at the pump to make up for the increased cost of doing business. As with all business, this cost will be passed through to the consumer or end user. That would add as much as another nickel to the price of gas at the pump. So who is paying that excess profits tax? It is the ignorant motorist who voted those incompetent oafs into office just to punish the oil companies. Cut off nose to spite face is the reality they ignore.

No matter when or how they are levied, taxes are ultimately paid by the final consumer. This is true of buyers of products or users of service. At the present time, hidden taxes amount to between 20% and 26% of every dollar the public pays for everything. Every dollar paid for every nail, hammer, car, vacation, legal service, doctor visit, and so on, now contains between twenty and twenty-six cents in hidden federal taxes. Actually, the hidden taxes on fuel are a bit lower than that because of the rapid rise of the cost of imported oil now around 70% of the price at the pump. A large part of that is appropriated by the governments of the oil-rich nations where we have no control. It goes directly into the coffers or certainly the complete control of their leaders. The socialist/communist welfare state will ultimately collapse economically when its natural resources run out, or it no longer has enough private capitalist enterprises generating profits to tax.

Currently and almost unbelievably, it is the capitalist machines of the Chinese state that are supporting the American socialist state systems by loaning that government vast amounts of money. What does the reader think will happen when the US owes China more than its net worth? With the trillions of dollars of debt we are now assuming, that time is rapidly approaching, thanks to the deliberate efforts of our current administration. Just who will be the ones to pay this enormous, outlandishly foolish debt?

When the American military suddenly finds itself paid, supplied, and owned by the Chinese and China takes over America in a bloodless coup, a group of Chinese bureaucrats will be running Washington for their own benefit. Will Americans wake up then?

Media and Environmentalist Promoted, Political Gold Mine, *Global Warming*

Since the use of systems and items described in this book would go a long way toward answering the demands of the global warming crowd, one would think I would be jumping on that bandwagon. If ever new information convinces me that global warming caused by carbon dioxide emissions from fossil fuels is real, and that it poses a serious threat, I will do so. At the present time, I believe all the hoopla about the dangers of global warming, are ill advised at best and could even be dangerously wrong. All humans act, react, make decisions and develop understanding and positions based on perception—not reality. It is not facts that motivate people, but their **perception** about everything. What one believes trumps knowledge in every instance, no matter how far from reality. So it is with *Global warming*.

While this book is not about climate change, it does reflect the effects of the political and media frenzies regarding *global warming*. Thus, the points are made that using alternative fuels or other energy sources could lessen or even stop the increase of carbon dioxide in the atmosphere, for whatever that may affect. A great many possible causes for changes in the average temperature of the earth are quite evident. The so called *greenhouse* effect of carbon dioxide, methane, water vapor and other gases is but one of a large number of factors that affect climate. Many of these factors are poorly understood. The actual weight of their effects on the climate is subject to large fluctuations, depending on which scientific study one reads. Even if it became a proven fact, global warming would be an insignificant problem compared with the real and present danger posed by rapidly rising oil prices and the resulting economic drain on our nation. Even the ever present threat of Islamic invasion and terrorism is a far greater menace than global warming at its worst.

The almost spiritual global warming movement is gaining large numbers of ardent and vocal followers. Many of these are blind disciples who have no clue about the realities of climate change, the physics of the atmospheric *greenhouse* gases or whether there is even the possibility of many of the claims put forth by the high priests of global warming. This has grown to universal acceptance as an absolute fact because it serves the political, social, cultural and/or economic agendas of its proponents.

Perhaps it is a present-day version of the Piltdown Man hoax, foisted off on unsuspecting scientists and the public almost a hundred years ago in 1912. That hoax took forty years to be completely discredited. In 1923 Franz Weidenreich, an anatomist, reported that the skull was a modern human cranium and the jaw was that of an orangutan whose teeth were filed down. It took scientists thirty years to concede that he was correct. Like most of us, scientists hate to admit error on their part. Many of us cling to dogmatic positions long after an error is discovered, and reality has become quite certain. Politicians and religious leaders in particular, are so infected. History provides countless examples like the murder of Huss and the imprisonment of Galileo. Some were extremely vicious, almost inhuman.

No global warming proponent ever acknowledges any of the following information. First of all, and most important, the term *greenhouse*, as applied to atmospheric gases, is a gross misnomer. The actual process whereby atmospheric gases retain heat energy and therefore cause the temperature of air to rise follows a very complex group of physical laws. These phenomena are distinctly different from what happens in an actual greenhouse. These laws involve the physical structure of the molecules of the various gases and how they resonate and/or rotate, when they absorb infrared radiation or heat. Each molecule both absorbs and emits radiation at different rates for different wavelengths and at different temperatures. This yields varying amounts of absorbed, radiated and retained heat energy. The only way we can measure these effects is to do so collectively using a significant amount of various gases, and a wide variety of mixtures of those gases including water vapor. A glance at just one report on the latest data research on this phenomenon tells us, "Recent improvements in the spectroscopic data for water vapor have significantly increased the near-infrared absorption in models of the earth's atmosphere." the full report is available at:

http://www.agu.org/pubs/crossref/2006/2005JD006796.shtml

Another report titled, *Water and Global Warming,* says, "Water is the main absorber of the sunlight in the atmosphere. The 13 million million (that's 13 trillion or 13,000,000,000,000!) tons of water in the atmosphere (~0.33% by weight) is responsible for about 70% of all atmospheric absorption of radiation, mainly in the infrared region where water shows strong absorption. It contributes significantly to the so called *greenhouse* effect ensuring a warm, habitable planet, but operates a negative feedback effect, due to cloud formation reflecting the sunlight away, to attenuate global warming. The water content of the atmosphere varies about 100-fold between the hot and humid tropics and the cold and dry polar ice deserts." The full article is available at:

http://www.lsbu.ac.uk/water/vibrat.html

There is another significant article on the effects of carbon dioxide at:

http://brneurosci.org/co2.html

Any global warming from the effects of carbon dioxide, if indeed it exists or poses any danger at all, is grossly distorted relative to the facts at hand. Most of the data used to show global warming are at best statistical and at worst, anecdotal. Both of which provide many opportunities for opinions (and agendas) to mitigate the resulting data. We know for certain that addition of any gas to the atmosphere, will contribute that gas's infrared absorption and radiation properties and all their complexities. All gases in the atmosphere have some *greenhouse* effect. This includes, nitrogen (75.0% - 78.08%), oxygen (20.11% - 20.95%), argon (0.89% - 0.93%), and carbon dioxide (0.035% - 0.038%). The percentages in parentheses are of air at sea level. Ranges are shown because air everywhere contains a variable amount of water vapor (from 1- 4% ±0.25%) and trace amounts of other gases. Each gas has a complex rate of infrared absorption, transmission, and emission at various infrared frequencies. Atmospheric water vapor content is from 20 to 120 times the amount of atmospheric carbon dioxide. It also has about 25 times the net temperature effect of the same amount of carbon dioxide, depending on various conditions. Bear in mind that net effect is the difference between energy from the Sun coming in, that heats the atmosphere, and energy from the atmosphere going out into space. Energy in is radiant energy from the sun being absorbed by atmospheric gasses and surface materials. It includes that convected from surface materials of the earth into the atmosphere. Energy out is that radiated from the surface that freely radiates into space and is not absorbed by those gasses on its way out through the atmosphere. It includes all energy emitted by atmospheric gases themselves out into space. Taking the varying amounts of each gas in the atmosphere into account results in a range for heat retention of water vapor between 500 and 3,000 times that of carbon dioxide. This number varies with temperature, altitude, location and water vapor content. All told, the effects of carbon dioxide on world climate is an extremely complex system with many variables. If all factors are considered in their proper proportions and even if the amount of carbon dioxide doubled, it would have a negligible effect on average global temperature.

The warmer air becomes, the more water vapor it can hold. Remember the weatherman's favorite *dew point* predictions? When the temperature lowers to that point, the air can hold no more water vapor, so it condenses out as *dew* or rain in the big picture. Using the same rationale as the global warming folks use for carbon dioxide, increasing amounts of water vapor would cause a much larger increase in atmospheric temperatures than carbon dioxide resulting in still warmer air and still more water vapor. Shouldn't this lead to a runaway greenhouse effect driving atmospheric temperatures higher and higher until the oceans boil and all life is extinguished? Obviously this has not happened, so something about these assumptions must be wrong for water vapor and carbon dioxide as well.

Another major factor water vapor adds to the mix is the heat of vaporization or condensation of water. Tremendous amounts of the sun's radiant energy is used to evaporate water all over the planet. All of that energy enters the atmosphere in water vapor. The warmer the ocean or wet land, the more energy goes into the air. When all this water vapor condenses out as rain, that energy is released and the air warms. This is the driving energy that causes the air to move and creates windstorms, tornados and hurricanes. For all practical purposes, the carbon dioxide content of the air has zero effect on the amount of energy that goes into the atmosphere or heats the air when water condenses.

One sizeable factor that man has affected tremendously is the water vapor that green plants give off. This is because of massive disruption in plant distribution by agriculture, land clearing, and lumbering. Our continuing decimation of all types of rain forests is removing a major source of water vapor that formerly entered the air. One example of this effect was used incorrectly as an example of global warming, which it was not. The disappearing snows of Mount Kilimanjaro are not an effect of global warming. Studies have shown that the cutting of the forest around the base of the mountain drastically reduced the amount of water vapor in the air flowing up the mountain. The result was that both the rainfall and snowfall on the higher slopes has also been reduced dramatically. This is one correct example of where human activity has interfered with nature. Deforestation worldwide has done far more damage to our environment and affected climate far more than even tripling the amount of carbon dioxide in the atmosphere could do. It alone could arguably be responsible for any temperature increase over the last hundred years. Reduction in the amount of water vapor would reduce cloud cover. This in turn would result in reflection of less of the sun's energy. Why don't we do something about that?

Whatever the effect of carbon dioxide, it is so small in comparison to many other factors as to raise questions about the real amount of the danger it poses. Certainly it is not the degree of danger claimed by the high priests of global warming. I seriously question the validity of the often quoted phrase, *Overwhelming numbers of scientists support the theory that man's use of fossil fuels is bringing about catastrophic global warming*. In the first place, the worldwide destructive clearing and burning of rain forest results in putting far more net carbon dioxide into the air than all the vehicles in the entire world. Second, shrinking rain forests mean less water vapor is released into the air. This could in turn mean less rain and snow where the air over land is drier. The questions remain, does the evaporation from the oceans increase and make up for this loss, and what effect does the drier air have on cloud cover and the resulting reflection of the sun's energy away from the earth? All these interacting variables have much larger net effects on global temperatures than carbon dioxide could possibly have. *Overwhelming numbers of scientists* may have no real clue about the degree of influence that carbon dioxide might have on atmospheric temperatures leading to global warming. Obviously it is much smaller than that of water vapor.

To explain why I make this statement, I have taken a small quotation from *The Creation* by E. O. Wilson. A national bestseller, the book was published by W. W. Norton & Company, Inc. of New York City. This quotation is about *scientists* and who they are.

Most researchers, including Nobel laureates, are narrow journeymen, with no more interest in the human condition than the usual run of laymen. Scientists are to science what masons are to cathedrals. Catch any one of them outside the workplace, and you would likely find someone leading an ordinary life preoccupied with quotidian tasks and pedestrian thought. Scientists seldom make leaps of imagination. Most, in fact, never truly have an original idea. Instead, they snuffle their way through masses of data and hypotheses (the latter are educated guesses to be tested—*global warming?*), sometimes excited but most of the time tranquil and easily distracted by corridor gossip and other entertainments. They have to be that way. The successful scientist thinks like a poet, and then only in rare moments of inspiration, if ever, and works like a bookkeeper the rest of the time. It is very hard to have an original thought. So for most of his career, the scientist is satisfied to enter the figures and balance the books.

Scientists are also like prospectors. Original discoveries are the gold and silver of their trade. If important, they can buy collegial prestige, and with it wider fame, royalties, and academic tenure. Scientists by and large are too modest to be prophets, too easily bored to be philosophers, and too trusting to be politicians. Lacking in street smarts, they are also easily fooled by confidence artists and sleight-of-hand tricksters. Never ask a scientist to test the claims of paranormal phenomena. Ask a professional magician.

Few scientists know more than a small fraction of available scientific knowledge, even within their own disciplines.

There is one certainty about the global warming movement. It has become a *cause celeb* and generated tremendous amounts of cash, mainly for politicians in the form of numerous and varied and punishing new taxes, cap and trade agreements, and expensive regulations. These taxes and the hundreds of global warming organizations constantly soliciting donations have turned it into a huge, self-perpetuating cash cow for its promoters and benefactors. This will guarantee its continuation long after it is proven untrue or at best, overblown far beyond any real danger.

In this writer's opinion, there are numerous other far more dangerous menaces facing humanity than global warming in its worst case scenario. Considering our burgeoning population combined with our steadily diminishing resources, one would think concern for this would be paramount in the minds of all thinking people. This is one of several score of serious menaces we are facing right now, albeit one of the most serious ones. While those major problems fester and grow with little comment or constructive effort, we continue concentrating our attention on things like global warming and the next American idol or soccer champion. Just as did Nero, the West fiddles as the world burns.

The following is an appropriate quote from Arnold Toynbee:

"We have been God-like in our planned breeding of our domesticated plants and animals, but we have been rabbit-like in our unplanned breeding of ourselves."

The text of this book takes no dogmatic position about carbon dioxide and global warming. Still, the author resents the *it's a proven fact* attitude of the many proponents of global warming, most of whom haven't a clue what the reality is. Its use as a money machine, a political club to bludgeon opponents, and an argument to influence the public, is quite shameful. The media-driven catchall assumption that carbon dioxide is the reason for global warming and that man's activities, particularly the use of fossil fuels, is the sole cause, is appalling. Many other factors that affect global climate change, can easily be shown to be much larger contributors to global warming than *greenhouse* gases. Many climate scientists even believe and report that we are near the beginning of another ice age. They certainly have just as much credibility as those with opposing views.

The disciples of global warming make one of their worst mistakes when they grossly simplify the effect of atmospheric carbon dioxide on world climate.

That is but a single factor among literally thousands in an infinitely complex set of interacting systems. We have been studying world climate a long time up to and including the age of the supercomputer and computer modeling. Still, we have hardly touched the surface as can be attested to by the accuracy of our current weather forecasting. For example, in spite of all our technology, predictions of the frequency, location, and path of any hurricane are fraught with pure conjecture. We can't even hope to predict the intensity of any hurricane season. Witness the 2006 season. Weather *experts* predicted it would be one of the worst ever. Instead, it turned out to be one of the mildest, the opposite of the predictions of our weather scientists and their supercomputer modeling. What about the reader's local weather forecaster? How often do those forecasts miss the mark predicting just a day ahead?

The world's climate system is vastly more complex than a single hurricane season. It moves in cycles and eddies that run from seconds to millennia. About forty years ago some climate pundits feared we were heading into global cooling and needed to prepare for a drier, cooler time with lower sea levels. According to many scientific studies of past frigid periods we are past due for the onset of the next ice age. Hubert Lamb of the UK Met Office dominated the 1961 UN meeting on global cooling. A founder of the Climate Research Unit at East Anglia, he was one of the world's top climate scientists. He warned that people had become complacent about climate at a time when population growth, cold, and drought could seriously damage their food supplies. (The Norse in Greenland perished of starvation after five hundred successful years when the Little Ice Age destroyed their crops.) In historic times the climate has veered from warmer than the present, the Medieval Warm Period, to the much colder conditions of the Little Ice Age. Evidence shows that much of the Sahara and the Middle East held lush vegetation and crop land tens of thousands of years ago. The media gleefully reported on global cooling as a fearful danger and asked, "What can we do about it?" Their tune has since changed substantially. The wind of media-driven opinion has switched direction from cooling to warming, with a vengeance.

Seemingly, members of the media are more interested in blaming Americans as the culprits causing the alleged problem than in trying to find those who may have viable solutions, or even reporting credible opposing views. It is a lot easier to blame others for a problem than to make the effort to find a real solution. That is one of our common human failings. In the media's case, it is far easier to find and report the negative emotional ravings than to ferret out and report the mundane realities of those seeking solutions. Pain and suffering, doom and gloom are their currency, their stock in trade. They thrive on the human *gawker* response that draws crowds of onlookers to even tragic accidents. Rarely do good news or real solutions to problems catch their attention. Many individuals the world over are using global warming as their new cash cow. Politicians are writing punitive tax legislation over carbon credits and charges. This practice is eagerly adopted on even small scales that tax ordinary individuals for tiny additions of carbon dioxide to the atmosphere. Politicians thrive on your money.

In spite of all this, there are several real happenings that do not support global warming. I recently visited Alaska and spent a day in Glacier Bay. While there I learned some interesting facts, mostly from a recent publication about the glaciers. Since the mid 1700s, Alaskan glaciers have been known to be steadily receding. Early explorers found glacier ice all the way to the mouth of what we now call Glacier Bay. There were maps in the book with lines showing the dates of glacier terminus from the 1700s to 2007. All the glaciers were shown to have steadily receded until the early 1990's. Since that time all of these glaciers have advanced steadily. In recent years, average global temperatures have dropped substantially. I also learned that arctic sea ice has been increasing rapidly since 2004. I wonder, why has the media not made the public aware of these facts?

The truth of the matter is that we are affecting the climate by adding carbon dioxide to the atmosphere. That's about the same kind of truth as the fact that pouring a bucket of water into lake Erie will raise the lake level. It is about the same order of rise in temperature that can be attributed to increased carbon dioxide in the atmosphere. The truth is we have little definitive knowledge of how much effect raising or even doubling the amount of carbon dioxide in the atmosphere will have in the long run. We are unable to hazard even so much as an intelligent guess as to what or how much the effect might be. Proponents of global warming neglect mentioning the following factors known to affect climate and the average world temperatures as much or more than increases in atmospheric carbon dioxide:

- The wobble of the earth's axis increases or decreases the retention of energy from the sun. (22,000-year cycle)

- The eccentricity of the earth's orbit increases or decreases the energy we receive from the sun. (12,000-year cycle)

- The variation of energy output by the sun. (1,400-year cycle)

- Variations in snow cover—snow reflects heat

- Variations in cloud cover—clouds reflect heat

- The variation in cosmic rays causes a variation in cloud cover. (no known cycle)

- Dust and sulphate in the air varies and can absorb or reflect more or less heat

- Ocean temperatures and circulation

- The destruction of forests—forests remove carbon dioxide from the air and release moisture

- Volcanic activity—eruption of Mount Pinatubo brought on several years of cooler temperatures.

- Winds—as winds increase, dust from dry farmland and deserts, enters the air (Gobi desert dust sometimes reaches as far as our west coast.)

It is important to know that the first three are all complex variables with many secondary effects on the temperature of the earth. They all are affected by the power of the *solar wind.* This powerful force affects our planet strongly and varies widely on an almost hourly basis. The solar wind is a stream of charged particles or plasma ejected from the upper atmosphere of the sun. It varies in its power and occurrence sometimes in short periods of time—days or even hours. Though the earth is protected from direct exposure to this energy by its magnetic field, some of this energy does reach the surface. How much it affects our atmosphere and climate is unknown. The noticeable effects include auroras, which are harmless, and magnetic or EMF disturbances, which can have devastating effects on electronic equipment. This includes computers, communication equipment, and navigation systems. These solar *storms* have even disrupted electric transmission shutting down large sections of the power grid. These forces can also strip away portions of our atmosphere forming a *tail* pointing away from the sun in much the same manner as a comet's tail. It is thought that Mars once had water and an atmosphere similar to of earth's, but it was mostly stripped away by solar wind over millions of years. Though the actual effect on the temperature of the earth is unknown, the solar wind could be a major and irregular modifier of atmospheric energy and thus climate.

The rest of the factors are also varying and can be interdependent yielding a complex mix of variables needed to produce any valid computer simulations. The effect of changes in the amount of carbon dioxide in the atmosphere is probably smaller than any of the other factors. In citing records from ancient ice cores that show the carbon dioxide content of air to be much smaller during previous ice ages, global warming proponents say this is proof that lower carbon dioxide levels cause lower air temperatures and higher carbon dioxide levels bring about higher air temperatures. Actually it is far more likely that temperature variations are the **cause** of changes in carbon dioxide levels rather than the result.

There is one gross misconception about ice and ocean levels that global warming proponents and the media constantly get wrong. It is true that water from melting glaciers and melting ice or snow supported by land will cause ocean levels to rise. However, polar ice and the floating ice shelves of Antarctica and Greenland do not do so no matter how big or extensive they are or how much they shrink. It's a simple law of physics known by most high school physics students. Floating ice displaces an equal weight of water so melting that ice will not change the water level whether it is in a glass, a bucket, or the ocean. Sadly, it seems that the proponents of global warming are more interested in the emotional impact of their statements than in their scientific accuracy.

For a level-headed and well-documented explanation of the realities of global warming/cooling factors, read the sections on climate in Nigel Calder's award winning book, *Magic Universe*. The information in this book is easily understood by laymen and is not influenced by political dogma.

The Real Reason for the Furor over Global Warming

It is the author's contention that the global warming issue has become a powerful force primarily because it provides a vehicle for politicians to gain revenue for their own purpose while giving them control over large industries and even nations. Using the powerful appeal of *save the planet,* these power hungry elitists and their willing supporters have generated an economic bonanza for themselves based almost entirely on emotional appeals poorly supported by facts. A check of the Internet will reveal several hundred web sites devoted to global warming. All of them are asking for members, donations to their *cause*, and participation. Most spend more words asking for donations than anything else. Global warming has become a gigantic source of funds for politicians and supporters of all kinds. I would like to know just how much money is collected by these *save the planet* web sites and how much they actually use to further their cause.

The East Anglia revelations: The pointed dishonesty of the weather *scientists* as revealed by the information *hacked* from university records is appalling. The truly sad and destructive inferences not missed by the public is that this throws a sinister shadow of doubt over the reports of the entire scientific community. The very integrity of those engaged in scientific research and reporting has been called into question everywhere. While honest mistakes and misjudgements have frequently caused doubts about *scientific* reports and publications, dishonesty has been an extremely minor issue. Now abject dishonesty including manipulation of computer code used in weather simulations on super computers has become evident in what is probably one of the most powerful *scientific* controversy of current times. This is without precedent and is deeply damaging to the entire scientific community. As a result, scientists will face an increasingly skeptical and doubting public for a very long time. The people have been lied to by a few scientists they trusted, and they certainly do not like it. Because of this, all of science has now come under suspicion, and that is a real shame.

About carbon dioxide: With the exception of hydrogen, all gaseous, liquid, and solid fuels produce carbon dioxide when burned in any energy process. In addition, the production of hydrogen by any means other than by electrolysis, using energy from nuclear, wind, water or tidal power plants will add carbon dioxide to the atmosphere from both the energy and the raw materials used to create the hydrogen (coal-fired power plants for instance). It is interesting to note that for each pound of carbon oxidized to carbon dioxide, four pounds of oxygen are removed from the atmosphere. For every thousand tons of carbon dioxide added to the atmosphere, eight hundred tons of oxygen are removed. In all the concern about carbon dioxide there has never been a single mention of that fact. This was pooh-poohed when called to the attention of one scientist who is an avid supporter of the anthropogenic global warming (AGW) movement. Sounds a bit like what the Soviets forced their scientists to report when facts conflicted with communist edicts. Our media ridiculed the Soviets when they did it. Now they are doing exactly the same thing themselves. My how times change.

It has long been accepted as fact that all of the oxygen in our atmosphere has been created by photosynthesis in plant life over hundreds of millions of years. Plants take carbon dioxide from the atmosphere, combine it with water to produce organic materials, and release oxygen as a byproduct. This has created a gigantic sink of carbon, including all fossil fuels as well as existing live plants and animals. It also built up the oxygen from zero to the 21 percent in today's atmosphere.

Human use of fossil fuels could be reversing that process by a tiny amount. This could arguably cause slight changes to natural, interactive processes, such as weather, ocean and air currents, water ice and sea levels. The truth of the matter is that we truly don't know how big the problem might be or even if it is a problem at all. It will be a very long time, probably centuries, before we have those answers.

Another seldom mentioned factor is the atmospheric gases including carbon dioxide that are dissolved or absorbed in the earth's oceans and surface waters. The warmer the water, the less of these gases are held in the oceans and surface waters. According to a report from the University of Hawaii:

"The ocean contains about sixty times more carbon in the form of dissolved inorganic carbon than in the pre-anthropogenic atmosphere. On time scales $<10^5$ years, the ocean is the largest inorganic carbon reservoir in exchange with atmospheric carbon dioxide (CO_2) and as a result, the ocean exerts a dominant control on atmospheric carbon dioxide levels."

This could easily explain why atmospheric levels of carbon dioxide dropped so low during the ice ages. At the time, the oceans absorbed much more of it as the equilibrium point moved to a much higher ratio of carbon dioxide in the oceans relative to that in the atmosphere driven by lower ocean temperatures. The movement of atmospheric carbon dioxide in and out of the oceans, driven by the water temperature, could be a big factor in determining the amount of carbon dioxide in the atmosphere. Since this process is not instantaneous, it could take hundreds of years for the balance to come to equilibrium. That is one more variable we must consider during air energy calculations.

Bully for Global Warming
and a Warmer Climate!

Wait a minute! Aren't the global warming fanatics looking at it backwards? Don't they have one very important fact upside down? Despite all that is being said on both sides, if it were real and its effect substantial, *Anthropogenic Global Warming*, could be the best thing to happen to our planet in many millennia and infinitely more good than bad. If they knew this, would the doomsayers of *global warming* ever mention it? Of course not. It would immediately derail their gravy train.

It is obvious that in the bigger picture and over the long-range, global warming would be quite advantageous for life on earth, even humans. The benefits would far outweigh the losses. It would certainly be a lot better than another ice age. Of that there is no doubt. So I say, quit wringing your hands and crying of doom. Global warming just could be one of the best things to happen since the last ice age ended.

Burn that oil. Burn that coal. Pump out that carbon dioxide! Let's heat up the planet and head for the beaches. How about palm trees in Labrador, oak forests in northern Alaska or a seaside villa on the warm shores of Greenland? Sounds pretty fantastic to me.

While there will certainly be many winners, there are bound to be a few losers. Sorry, New York and LA, but you'll have to move a bit inland like quite a few other coastal habitats. I'm terribly sorry about Florida, but no change is ever all good. Besides, the lush semi tropical shores of Labrador will be habitat for former Florida flora and fauna while the lowlands of Georgia, the Carolinas, and the rest of the southern states all the way to the West Coast will be lush tropical paradises. That's certainly a much prettier picture than that painted by the doom sayers of the church of **horrible** global warming. I'd certainly like to hear a response from some of their members. Al, eat your heart out!

The one glaring error the believers in **detrimental** *global warming* make is that it is **detrimental**. The fact is it would be almost entirely **beneficial** to mankind and to all life on the planet, and substantially so. Warm periods, such as the Medieval Climate Optimum, were historically beneficial for civilization. In contrast, corresponding cooling events, such as the

Little Ice Age, were uniformly bad news. Up to a point, the warmer our planet gets, the more life it supports and the more variety that life has. In contrast, the colder it gets, the less life and the fewer species exist. All fossil records support that reality as far back as the Permian period, 500 million + years ago.

We have very good evidence that a warmer planet has far more benefits than negatives. For obvious reasons, *global warming* advocates never consider that very significant fact. It is well known from several sciences including botany, zoology, and paleontology that most all life has fared much better during warm periods on the planet than during colder periods. The amount and diversity of life were severely diminished during the ice ages, expanded as the glaciers receded, and were much greater during periods warmer than the present. During warmer periods, the borders between tropical weather and temperate, as well as temperate and arctic on our planet, move toward the polar regions. As a result, the quantity and variety of life increases when temperate areas gave way to tropical, and arctic gave way to temperate. This is a well-established and easily understood fact.

Just compare the number and variety of life as it now exists in the tropics, in the temperate zone, and in the arctic. Imagine what would happen if the tropics expanded and the arctic zone shrunk. Would that be beneficial to life or detrimental? Obviously it would be beneficial. I'll wager few people ever even considered that fact. There is, in fact, a greater number and variety of species in a square mile of Amazon jungle than in the entire Arctic and Antarctic regions of our planet. For obvious reasons, the high priests of the harmful global warming movement would never mention this fact.

A case in point in human experience: The Vikings established a thriving community of farms, villages and churches in Greenland that lasted for at least 500 years during the medieval warm period that was somewhat warmer than today. There is even evidence they established colonies in North America (Vinland?) during the same period. Then, when the *little ice age* happened and arctic weather moved south over Greenland, they starved and disappeared leaving their fields, farm buildings, homes and churches where they stood. Greenland's tiny Inuit population persisted until the present. In the eighteenth century, the Danes reestablished a small colony where one of the Viking settlements had been. That colony has survived only with imports of substantial amounts of food, building materials, and hay for animals. For all practical purposes, the Vikings are still the only self-sustaining population of Greenland, other than a few Inuit hunters. Present-day greenlanders now place their hopes for the future on utilization of vast, untapped mineral resources

Suppose the worst claimed by the gurus of *global warming* actually happened and sea levels rose a hundred feet. During this period, the tropics would expand 500 to 1,000 miles, and the temperate zones would move toward the poles about the same distance, Should this happen, large areas of northern Canada and Siberia, among others, would become much more hospitable to life of many kinds—temperate forests and fertile farmlands. Since it is well known that warmer periods brought wetter weather patterns, many deserts would become lush and green. Talk about a green revolution—that would be a real one! Southern California, Arizona, New Mexico and West Texas valleys could become tropical jungles. The same could happen to the deserts: the Sahara, the Gobi, and even the Australian. Wouldn't that be a kick? Who knows, with the ice gone, Greenland might even become green and habitable. One thing for sure, the Arctic ocean would become an open shipping lane. The overall benefit to life on the planet would be huge. So we would lose some land and a few cities along the coast and many islands would shrink or disappear, big deal! Since it would happen gradually, the people and much of their property could easily be moved elsewhere.

Oh yes, the polar bears and penguins. Those cute little birds and those big, soft and cuddly bears so painted by the brush of the media to appeal to children. Actually, they are both predators (as are we humans, even children). Penguins kill and eat millions of poor little fishes. And polar bears? They kill and eat many seal pups every year. Just think of those poor, cute little seal pups being brutally dragged from their lairs, slaughtered, and eaten. Also, they would gladly kill and eat you if given the opportunity and they were hungry. Of course, they much prefer tasty seal pups. Should the planet warm substantially, the environment of their habitat would change and they would probably go extinct. Sad, it not?

The expanded tropics would become habitats for many individuals of hundreds of new species, many just as loveable as penguins and polar bears. Think more cute little monkeys and colorful new birds in lush jungles teeming with life. Think also about the manatees, armadillos, big cats, alligators, crocodiles, snakes, lizards, spiders, and thousands of other creatures you never see in the arctic. They would be found in countless numbers in temperate areas that replace arctic tundra and permafrost, as well as in the tropics that replace temperate areas. Compare those habitats with the cold, bleak, windswept arctic of snow and ice, supporting but a few creatures in hundreds of square miles.

And how about this for you or me, a sunny, warm beach with palm trees and colorful birds compared with a frigid, windswept snowy plain, or maybe even farmland on ground once covered with a mile of ice in Greenland. That same ice once covered almost all of Canada and most of the northern United States and Europe. Global warming removed that ice and turned that land into woods, farms, vineyards, and varied wild habitat? Considering those known and obvious facts, does global warming seem so terrible?

Remember those foreboding warnings about nuclear winter, with the cold killing everything? How about the same thing caused by volcanic eruptions or an enormous meteor strike? Contrast that with a carbon dioxide summer, where the arctic ice melts and everything gets warmer. Palm trees on Labrador, temperate forests on Greenland and tropical rivers crossing the Sahara. Long-range, that is what global warming would produce. It has in the past, long before man existed, and could certainly do so again in the future, probably long after man has gone extinct.

Another shocker the government controlled media will not report relates to new information on the benefits of increased atmospheric carbon dioxide. This quote comes from a scientist and professor from a well-known university whose specialty is studying the effects on plant life of increased carbon dioxide in the atmosphere. It was a comment made to a blog I posted of this *Bully for Global Warming* chapter. He does not want his identity revealed for obvious reasons. (Peer pressure and access to government grant money.) He wrote:

"What your very good essay doesn't much focus on is the tremendous increase in plant growth that has already been afforded by the increase in carbon dioxide. This is generally calculated to be in the range of 15% to 40%, depending on the species (and whether it's a C4 or C3 type of photosynthesis) – which means at minimum about a 1/7th increase in plant productivity. This means in turn that something like a billion humans are being fed by the EXTRA crop growth from carbon dioxide. So the contributions of Man, through fossil fuel, land use changes, and agriculture, have a very positive effect. Just think of the harm – the starvation -- if the extra carbon dioxide could be instantly made to go away, as the catastrophists apparently wish. It is not a pretty scenario at all."

After realizing these facts, maybe Al Gore will write a new book titled, *A Beautiful Truth* that will tell the facts about global warming. Oh, but that just wouldn't work with his liberal agenda, would it? The new American liberal politics trumps facts, science, and all other forms of truth. Shades of the old Soviet Union. This is especially true of all information showing the benefits of increased carbon dioxide and of global warming, whether related or not. Certainly, the degree of their relationship has not been established by anything other than anecdotal evidence.

Some Thoughts about Other Realities

Smokescreen Hate Campaigns Against *Big Oil*

Hatred of *Big Oil* promulgated by leftist politicians and the media, is a deliberate tactic intended to inflame the ignorant public and take the spotlight off the gross failures of government efforts in virtually every government-run project, especially energy. This and other similarly imaginary factors add to the price and subtract from the available supply of petroleum, in addition to the real problem of diminishing oil reserves.

Thanks to some environmental extremists and their liberal supporters in Congress, we have virtually castrated oil exploration and drilling for new sources in and around the U. S. These powerful voices have made it all but impossible to build new refineries and have even tried to stop the expansion of existing refineries. The constant stream of condemnation of the oil industry has convinced the public and thus many politicians that Big Oil is their enemy. This precipitated tremendous animosities that are amplified every time the highly advertised price of gasoline goes up. The government's initiatives for promoting alternative fuels and limiting consumption have put a damper on nearly all new investment in infrastructure to increase production and capacity by oil companies. Our country is awash in a virtual sea of proven oil fields from the Gulf of Mexico to the North Slope. This oil is there and accessible, but in many areas where it is relatively easy to extract, American oil companies are forbidden by law to go after it. As a result, they must go where drilling is much more difficult and dangerous, BP in the deepwater of the Gulf for example. No perpetrators of the problem, American oil companies are almost as much a victim of high-priced petroleum as the average motorist. Certainly their image is tarnished by all the class hatred rhetoric.

If anyone is to blame it is the extremism of over zealous environmentalists who have helped drive the price of fuel through the roof. They seem to want to make many things, and especially fuel, so expensive we have to curtail its use or stop using it altogether. They want us all to be riding bicycles or walking to satisfy their warped sense of necessity. Think about American oil companies being forbidden to drill for proven oil deposits around Florida to protect shoreline ecosystems. Does it not sound reasonable—protecting the environment? Now that we are not drilling in these fields, the world's largest oil company, PetroChina in cooperation with Cuba is drilling not far from Key West where American companies are forbidden to drill. Ask any environmental activist what they plan to do about that. PetroChina has none of the environmental controls on drilling and drilling techniques required of American oil companies. So much for our politicians protecting the environment along our shores. As far as Big Oil is concerned, last year, PetroChina overtook Exxon as the world's largest publicly traded oil company. There goes all that potential profit and billions in taxes from our shores to China, not to mention the thousands of good jobs that left our shores. Real smart move, Democrats. An enemy bent on destroying our economy could hardly have accomplished more.

All this posturing and the blaming of those deemed responsible for rising fuel prices are so much useless and potentially damaging hot air. The fact is world demand for fuel is growing while the supply is diminishing despite the momentary drop in oil prices brought about by the worldwide recession. No matter how experts consider it, that's why oil prices will rise in the long term. New oil fields are becoming harder and more expensive to discover and utilize worldwide.

Here are some facts. Some analysts say high oil prices and the record profits they create, are masking growing difficulties at many of the major Western oil giants. American oil companies, which once dominated the global energy business, now control only about 6 percent of the world's oil reserves. They are having a difficult time increasing production and renewing reserves. They are being replaced by resurgent national oil companies such as PetroChina, Brazil's Petrobras, Petroleos de Venezuela, S.A. and Russia's Gazprom. Politicians supported by the media do not seem to care a bit about that as they keep screaming hate at American oil companies and inciting animosity among ignorant voters. If you think fuel prices are high now, consider what could happen if taxes and government animosity drive American oil companies out of business or cause the stockholders to merge them with foreign companies.

That has happened to other industries in the past. Textiles and consumer electronics are the first to come to mind. In their lust for power, some politicians seem bent on using class envy to provide them the political power to kill every American golden egg-laying goose they can. This economic suicide seems to serve no purpose other than use the poor and ignorant to bring down the very people who have built the most powerful machine for innovation and wealth for the highest number of common people the world has ever seen. Will they do the same to any alternate fuel or energy system that starts to show promise? It wouldn't surprise me at all. They seem to want every promising solution thwarted or stopped. Perhaps this is because they do not want any solutions. As long as there are problems they can use as clubs with which to beat their opponents bloody, they will continue to do so. Is this all because their lack of ability to build has turned them to doing what any idiot can do, destroy?

It was Abraham Lincoln, champion of the poor and downtrodden, who wanted no reprisals against southern leaders after the war. Lincoln felt it best to turn enemies into friends. During the Civil War, he refused to demonize the enemy. When Washington received news Robert E. Lee had surrendered at Appomattox, the President was asked to speak to the celebrating crowd.

Addressing the jubilant crowd, Lincoln said,

"I have always thought Dixie' one of the best tunes I have ever heard. Our adversaries over the way attempted to appropriate it, but I insisted that we fairly captured it. I presented it to the Attorney General, and he gave it as his legal opinion that it is our lawful prize. I now request the band to favor me with its performance."

The band played *Dixie*, followed by *Yankee Doodle*. This was Lincoln the humanitarian who in his second inaugural speech finished it:

"With malice toward none. With charity for all."

Everyone around Lincoln knew that he meant it. How many of the current crop of politicians could say and mean words like that?

In stark and obvious contrast, Democrats, like DNC chairman Howard Dean, are diametrically opposed to such thoughts. His famous rant, "I hate Republicans! I hate conservatives! I hate Rush Limbaugh!" gives ample proof. Hatred and demonization are the political stock-in-trade of the left including virtually all current liberal Democrats as well as their vocal friends and supporters. They would have hated Lincoln as well.

So much for peaceful cooperation.

The Advantages of Size

In the automotive world, Joe Doaks and company might be able to produce only a few cars a week after three years of hard work simply because this small company has limited working capital. They must then sell the cars to get more working capital to continue in business. The profits from sales must be enough to pay the employees, pay for raw materials, business services, insurance, legal fees, a location where the business can operate, and taxes and fees to several governments. Those same profits must pay for any expansion and for interest on any debt the company may have.

On the other hand, Chrysler, General Motors, or Ford already have the needed physical assets and capital to pay for the development and manufacture of any new vehicle they deem will make their sales goals and earn a profit. They also have legal and contractual obligations that Joe Daoks and company do not. While development and all the associated details may take any of the *big three* eight years to complete for a single vehicle, once those goals are met they have the ability to produce these new cars by the thousands. This enables them to swamp

JD and company with saleable products. Of course, if JD and company have a valid patent on a critical part of the new car, even GM couldn't produce it. Usually what happens is that GM or another big auto company will buy the patent from JD and Company for more money than the owners would make on their own.

Government Run Enterprises

Now that the government is running GM for the benefit of the UAW it is doubtful any truly innovative products will be forthcoming from their research. With profits no longer the driving force for GM, having been replaced by social services for the UAW, the ultimate demise of a once great American corporation is assured. While the innovators and genius that powered this automotive legend move to other employers, GM will shrink until it disappears. That shrinkage has already progressed a long way toward the time when all that will remain will be empty factories, unemployed UAW members, a diminishing number of parts suppliers to keep GM products running for a while, and US taxpayers paying for a mountain of employee payroll and benefit debt. Foreign auto makers will quickly fill in the void. This is probably the best thing that could happen to Ford Motor Company. Their sales have grown substantially while those of GM (Government Motors) and Chrysler have slumped. The author personally would never buy a car made by a company run by the government or the UAW.

Detroit and Michigan are prime examples of what liberal politics and union dominated industries can do to a city and state. No words can counter the power of a view of the wasteland of abandoned houses and empty, refuse-filled lots that once was a vibrant city. It is the remains of a war zone where the people lost. It demonstrates clearly what the policies and efforts of the left can do. It is where corrupt politicians and union goons created their utopia. Is our entire nation headed in the same direction? Unfortunately, it looks that way.

Government has many times proven its inability to run any enterprise successfully. Witness Amtrak, the Postal Service, Social Security, Medicare, and the list goes on and on. All are bankrupt or soon will be. They require vast sums of taxpayer money to support their inefficient operation. Can the reader think of any government run enterprise that has succeeded in doing anything other than add to the bankruptcy of our nation?

Private, capitalist enterprise with its multi lithic structure, wide spread risk, profit incentive, and encouragement for innovators is far superior to government in every enterprise. This includes providing the highest standard of living and the most widely spread benefits to all of the people. It also includes far more of virtually every desired end the left say they want than any socialist/communist government has ever achieved. Every socialist/communist state eventually degenerates into a political reality where a few elite individuals control the government and the means of creating wealth. It is where the vast majority of the people become wards of the state spending their lives serving the state for the benefit of the elite group that hold all the power. There are few exceptions. Entrepreneurial innovation and creativity virtually die in state run systems.

These same sorely needed creative forces flourish in the competitive and inspirational atmosphere of capitalism. Even the Chinese Communist government finally realized that. It became obvious when they found individual farmers on their tiny personal plots were far out-producing the giant government collective farms. By embracing capitalism, the Chinese have turned into a virtual success machine. The wealth even their limited form of capitalism has produced is astounding, and in an extremely short time. The standard of living of most Chinese has already risen far above the best their strictly controlled economy ever produced.

The concerted effort to solve our energy problems would certainly stimulate the economic growth that has sustained our economy at such high levels for so long, if augmented by this kind of positive attitude and action by the general public. If hope for substantial economic rewards to individual entrepreneurs drives our efforts at solving our energy crisis, we will succeed and hugely so. If on the other hand, those anti business voices of doom and gloom succeed and control our government with the new, oppressive regulations and taxes they have promised, our economy will go into reverse and much more quickly than even the present slow down has indicated. How can they promise *half a million new jobs* when they also promise huge tax increases on those who would otherwise provide the capital to create those jobs? More than ninety percent of the *new* jobs created since the liberal Democrats took control, are government bureaucrat jobs that soak up more taxpayer money. Faced with the promised tax increases on business (the wealthy), it is more likely that we will lose several million existing jobs from the private sector along with the new ones. The fall of the dollar we are currently experiencing will accelerate. Those entrepreneurs who might have solved our energy crisis will do so in Ireland, or China, or India as our stifled economy sinks into depression, and our energy needs go unsolved.

The recent crane accidents are a significant and troubling indication that this may already be underway. This may seem a remote relationship, but consider: construction cranes are disassembled after their job is finished and shipped to a new site for a new job often far from the last site. The largest number of construction cranes in use, long was in the U.S. Now it is China and India that lead as locations for construction cranes with the U.S. a distant third, almost on a par with Russia. The result is that the newest and best cranes are occupied elsewhere, and we have to be satisfied with the left overs, sometimes aged and second rate.

Cranes, construction equipment, investment capital and even people with expertise now move in a world economy. These important wealth generating systems and people gravitate to the locations where they have the most opportunity to generate wealth. They will abandon places where government regulations or taxes reduce the wealth they can produce and move to more favorable environments. Understanding this brought about the explosive Irish economy and even the new Chinese economic expansion. One example of this in action is the movement of some American managers and craftsmen out of the U.S. Russian builders are recruiting construction managers, contractors, and even skilled workers to work in Russia. This is a new thing, and quite troubling. It shows the relative energy of those economies. I'm sure there are other similar movements of people, investment capital, and technologies in this world economy. People, equipment, technology and money are now moving worldwide to where they can generate the most income. It is interesting to note that last year more than two million Americans moved to other countries, mostly for better work opportunities. The previous high, just a year earlier, was half of that.

How we approach and deal with this serious problem, and the attitude we take toward those who have the power to solve it, will ultimately decide which technologies prosper and which fall by the wayside. The steadily rising cost of petroleum fuels has made alternative systems practical that were far too expensive when oil was five dollars a barrel. One possible stumbling block to these changes could result from a dramatic, sudden, and sustained drop in the price of oil like that resulting from the recession of 2008. As long as these low prices continue, development of alternate fuels and energy systems will probably be put on the back burner. Eventually, dwindling supplies of petroleum will wipe this out as demand outpaces supply and prices head skyward once again. Solving the energy crisis will impact many serious problems that we are facing right now. The biggest is the outflow of billions of dollars for oil to nations that preach our destruction. The boost to our economy alone would create a bonanza in this country like we have never before seen. New jobs, new technologies, new industries, and new entrepreneurs would flourish. Even without new taxes (what a dream that is) government revenue would soar from the increased economic activity. The demise of the oil industry would certainly be replaced by the new energy industries. Those oil companies that get aboard these new technologies rather than opposing them, could use their present wealth to invest in them and grow rather than fade away.

Whether or not human contribution of carbon dioxide to the atmosphere causes global warming is only remotely connected to our energy related problems. If these new energy systems eliminate the wholesale use of fossil fuels, so much the better. Even if it has a negligible effect on global warming, it can do no harm to maintain the level of carbon dioxide in the atmosphere near to where it has been for a long time. That would also satisfy the demands of the global warming proponents and opponents and redirect their energies elsewhere to the other, far more dangerous challenges facing humanity. Concrete benefits to the rest of the world with the exception of the oil-rich nations would also be substantial.

A Wild (and extremely unlikely) Future Scenario

This fanciful scenario is extremely unlikely, but it could put America back on track to returning to the most free, most successful, most innovative, wealthiest nation on earth. This would include that wealth spread widely among a grateful populace with virtually unlimited opportunity for every individual to lead a full and rewarding life. Many in the free world might follow our lead and move into an era of unprecedented prosperity while saving our planet from the environmental destruction now running rampant. I could list the many destructive actions now consuming our natural world, but suffice to say human creativity could overcome these destructive actions by peaceful means. Here starts that scenario:

Nov. 7, 2012—Associated Press: (hypothetical)

The American presidential election is over with the results a foregone conclusion. Barack Obama won by the largest majority popular vote in history. He won every state in the Union while his two opponents didn't come close to capturing a single electoral vote. How did this huge reversal of public opinion come about?

In the spring of 2010, his programs were failing everywhere. The recession was deepening, job losses continued to mount, and unemployment rose past 12%. Even with the new taxes, federal revenue was dropping rapidly. To make matters worse, the Chinese and several other creditor nations were threatening to shut off our credit. The reverberations of this were felt in Wall street, main street, and even the alleyways of inner city America. Obama's approval numbers were plummeting and Congress was in an almost uncontrollable uproar while their public confidence was in free fall.

During this troubling period, Obama spent time studying history and the lives of historic figures for answers. He claims to have had a sudden realization while studying the creation of our Constitution and the early years of our fledgling government. His comment, "I have had it all wrong. Socialism and government are not the panaceas I believed them to be." was startling to say the least. In September of 2010 before both houses of Congress, he made the following speech:

I have looked carefully at those nations around the world who have exhibited the best and biggest growth of wealth of their people. Those nations, not who had great wealth, but those whose populace was showing the largest increase of personal wealth and well being from wherever they started improving. I did not concern myself with where they were at present, but instead concentrated on those showing the biggest improvement in the welfare of the individual and the most promise for continued improvement. The nations I studied were China, India, Ireland, Brazil, and even post WWII America.

I was impressed with a number of things. The most important information came from both China and Russia. Officials in both Communist nations noticed individual peasant farmers, working their own, tiny, private plots of land as gardens were outproducing the collective farms in both quantity and quality of produce by a huge factor. At markets, their produce was the first choice of shoppers. No one touched the produce of the collectives until that offered by the individual farmers was gone. How could this be, individuals besting the state by so much, and in their precious spare time? That certainly was not Marxist.

Noticing this, both Chinese and Russian government officials came to realize one of their basic tenets was dead wrong. Those who realized it first soon began allowing those individual farmers the time and land to cultivate bigger and bigger crops and to supply their surplus to that of the regions' collectives. They were paid well for what they produced and began to prosper more than those who worked on the collectives. They had, in fact, become capitalists, working hard and caring for the land and what it produced.

As news of this *revolutionary* practice spread and reached the highest levels of government it wasn't long before more and more *capitalist* enterprises, even manufacturing, began contributing to the Chinese economy. The Communist government soon embraced limited capitalism and the phenomenal growth of the Chinese economy was off and running. It reached the point where Tom Friedman in his book, *The World is flat,* wrote, 'Their current leader, a scientist and engineer reportedly announced, *Profits are good! Business is good! Free-enterprise is good! Capitalism is good!*' This was an amazing reversal of political form, especially for the Chinese Communists. The same type of thing was happening in Russia and even India although not as phenomenally as in China.

I then took a look at what had happened in Ireland. In a few decades, Ireland moved from a poverty-stricken, virtually third-world nation to the most dynamic economy in Europe. They produced so many jobs they had to import workers from all over the world.

The Irish went from one of the most poorly paid workforces in Europe to one of the most highly paid in about twenty years. This new wealth was fairly evenly distributed among all the people. How did they do it? They dropped their business income tax rate first to 20%, then to 12.5% including capital gains. Not only that, but the government adopted a probusiness attitude and profits became the good, positive word it used to be. This was being echoed in the media and the attitude of the people soon began to change.

If you were an investor would you move your money to the U.S. with a promised 40% to 50% corporate tax and a 39% capital gains tax, or Ireland with a 12.5% corporate tax that includes capital gains? It's no wonder the US stock market plummeted. Investment money fled our shores for better returns elsewhere. Also, talented people began moving from our shores to other countries with more business freedom and better opportunities. In 2008 there were twice as many Americans deserting our country for work elsewhere than ever before.

And how about post WWII America? There were race riots and union riots, company goons and union goons, and terrible inequalities that took place over many years. These have been overcome virtually entirely. There were also lots of very decent people, people of all races, working together to ease the racial tensions and move toward true equality including individual civil rights. Sure there is still racism and racists. I have never believed racism is limited to Caucasians. People of color are also guilty of racism and regularly use it to serve their own selfish desires. Racial, religious, political, cultural, ethnic, and other differences can create conflicts, but certainly they can and must be overcome. All of our ancestors and even some of us here and now have been guilty of terrible crimes and cruelty against others. Let us all realize the past cannot be changed and work together for a better future for all of us and for our children. Let us not open up old wounds, but soothe them and deal in the present and prepare for the future. Let us abandon the *get even* attitude that seems such a powerful, yet unrealistic, human trait.

What have I learned from all of this? Why has it brought me to completely reverse my thinking and efforts? First of all, I came to realize that thousands of companies, groups of people, working for an organized purpose and free to employ other free people, will do a far better job of running the economy than a single government entity. Government's job should be to see to it that laws are created and enforced to prevent specific abuses of power in both the private and the public sector. Government should not be in the business of running any enterprise that can be handled by the private sector.

Excessive taxes have been proven many times over to be a serious deterrent to business and profitability and so to job growth. Profits generate jobs and high taxes result in low profits, poor job growth and actually decreased revenue. Tax cuts for business have almost always resulted in revenue growth while tax increases have always resulted in less revenue. Ireland's economic growth took off when business taxes including capital gains were reduced substantially. From around 40 percent to 20 percent. This was so successful and created a business expansion that actually doubled the total tax revenue to the government. As a result the business income tax and capital gains tax were cut again to 12.5 percent. Once more total tax revenue increased and companies from all over the world began opening divisions in Ireland. This brought world class banking and other services creating even more growth. The Chinese and other progressive nations have adopted similar tax policies with similar results.

I thought carefully of what this could mean to my country. How could we adopt such a program without depriving the poor and giving a huge benefit to the *wealthy* for whom I had so long expressed such opposition? Then I envisioned a program that would serve both purposes, reward success without granting economic power to a few wealthy individuals, and offer real assistance for the poor by providing job training for those willing and able to work. This would in turn enable them to work their way into better economic conditions. This realization changed my mind about the direction our nation should be taking.

As a result of these realizations I am proposing the following be enacted as soon as possible:

1. Business income taxes to be reduced to 20% for all businesses.
2. Business income taxes for those engaged in a detailed list of environmentally sound, energy efficient development be dropped to 12%
3. Business income taxes for those engaged in a much shorter list of high priority development be suspended entirely for a period of ten years.
4. The present complex personal income tax system be scrapped and replaced by a far simpler system. This new system would include a basic flat tax, a transfer tax that would apply to all kinds of property transfers for any reason, a negative income tax to replace federal welfare, and the removal of all other federal taxes, hidden and otherwise.
5. All American military bases on foreign soil that are not directly involved in combat or combat support are to be closed and the military equipment and personnel returned to US soil.

The details of these new proposals are contained in the complete document I am now presenting to Congress. The resulting shrinkage of government payrolls would offer substantial savings and those employees laid off would be given training, so they could enter the private workforce in a growing economy. My ultimate goal would be to cut government payrolls to less than half of the present.

Needless to say, Congress, the news media, and the entire populace were stunned. The proposals he then put before Congress were earth-shattering. A new era of *change* had burst upon the scene with unbelievable results.

It took nearly six months of hard work to get President Obama's proposals written into law. At first Congress balked, but the public clamor and outcry supporting the proposals combined with the pressure of an upcoming midterm election, eventually ended in the President signing into law the most comprehensive change in American government since the Constitution.

Midterm elections that followed dethroned many diehard leftists who continued to oppose the changes in word and deed. The surprising thing about the elections was the removal of so many long term members of both houses. Those who favored and supported the new laws were elected handily over the very weak opposition. Though Congress remained fairly evenly split between Republicans and Democrats, there was no doubt the vast majority were strong supporters of the new Obama proposals.

It didn't take long for the economic engine of America to spring into action. With business taxes lowered and the government's new attitude toward profits, investment capital flowed into the energy and transportation sectors creating many good jobs. This was soon followed by growing employment in virtually all sectors. By the end of 2011, unemployment had dropped substantially and public confidence had skyrocketed. The dollar was now rising on world markets, and the long sluggish American stock market began a steady improvement. World economic conditions even improved, spurred by increased activity in America. The environmentalists were mostly pleased because the expansion was largely in what could be dubbed *green* favoring areas of the energy sector.

By the time campaigning for the elections of 2012 was well underway it was clear that President Obama's new initiatives were so popular nearly every potential candidate was on the bandwagon. Republicans, Democrats, Libertarians, and Independents, were all praising the new order and clutching for Obama's coattails. Virtually no one could speak out in opposition. Those that did were frequently panned by a much changed media. Even the class warfare, so prevalent a few years back was toned down considerably. Success filled every dream and ethnic, racial, and other animosities virtually disappeared but for a few sore heads and political fundamentalists. Political commentators across the board from all points of view were incredulous. Even Rush Limbaugh was applauding Obama, an unbelievable happening. The result of the upcoming Presidential election was a certainty. In the words of several reporters, "Obama will be reelected by acclamation."

End of scenario

I know, that's a foolish dream with little basis in reality, but wouldn't most Americans wish it could come true? Wouldn't the vast majority of Americans be much better off? The only ones to suffer would be those self-serving politicians who are only interested in money and power and will do virtually anything to get it or hold on to it. I doubt many tears would be shed for them or their followers. Maybe we will be fortunate, and another president will do what was described in this fantasy after he replaces Obama when he gets thrown out of office in 2012. It is plain the author cannot see Obama changing his colors.

SECTION V

References and Recommendations

Some author recommended books include the following:

Calder, Nigel, *Magic Universe,* Oxford University Press.

Diamond, Jared, *Collapse,* Viking Penguin Books.
_____ *Guns, Germs and Steel*, W. W. Norton & Company, Pulitzer prize winner.
_____ *The Third Chimpanzee,* Harper Collins.
_____ *Evolution and the Future of the Human Animal*, Harper Collins.

Friedman, Thomas L., *The World is Flat Rel 2.0,* Farrar, Straus and Giroux.
_____ *The World is Hot, Flat, and Crowded,* Farrar, Straus and Giroux.

Gladwell, Malcolm, *The Tipping Point*, Back Bay Books.

Gore, Al, *An Inconvenient Truth,* Rodale Publishing.

Gould, Stephen J., *Bully for Brontosaurus,* W. W. Norton and Company.

Hoffer, Eric, *The True Believer,* Harper and Rowe, Publishers.
_____ *The Temper of our Times,* Harper and Rowe, Publishers.

Stossel, John *Give Me a Break*, Harper Collins - September 2005.
_____ *Myths, Lies, and Downright Stupidity*, Harper Collins May 2006.

Bibliography

Capstone Turbine Corporation (Nasdaq: CPST), NEWS release, *Diesel Fuel News*, June 11, 2001.

_____ NEWS release Source: Ventura County Star, June 7, 2001.

Friedman, Thomas L., *The World Is Flat Rel 2.0,* Farrar, Straus and Giroux.

_____ *The World is Flat, Hot, and Crowded,* Farrar, Straus and Giroux.

Roan, Vernon, Principal Investigator, **Daniel Betts, Amy Twining, Khiem Dinh, Paul Wassink and Timothy Simmons,** *Final Report—An Investigation of the Feasibility of Coal-based Methanol for Application in Transportation Fuel-cell* SYSTEMS, University of Florida, Gainesville, Florida, April 2004.

Stauffer, Nancy, Report, *MIT Laboratory for Energy and the Environment,* Boston, Mar 07, 2003.

UCI's National Fuel cell Research Center, *Report on new, efficient and clean power generation technology,* Irvine, Calif., June 15, 1999.

Wald, Matthew L., *Hydrogen Fuel cell Promise, Scientific American*, May 2004 issue.

Internet references and links

NOTE: Internet sites come and go over time, many lasting only a few months. Some of the specific sites listed in this book are already gone as it goes to press. If a site as listed is no longer available, reduce the end section of the address and try again. Go to the .com, .net, .edu part of the address and search there. Later reports or articles on the same subject have often replaced the original one listed.

~~~~~~

NOTE: all autobloggreen.com articles are removed after about a year. Goto: **http://www.autobloggreen.com** and search for current articles

Aptera electric car information:

http://www.autobloggreen.com/category/aptera/

Capstone Turbine Corp., New Solutions: **http://www.capstoneturbine.com**

Go Green with your home or business. The latest products designed to save you energy and reduce your energy costs. To find out how, go to:

http://www.ecoenergyanswers.com

Remy International, Pendelton, Indiana - manufacturers of HVH electric motors for hybrids and EVs:

http://www.remyinc.com/index.asp

UCI National Fuel cell Research Center at UC Report on micro turbine tests:

http://today.uci.edu/news/release_detail.asp?key=694

And for the political realities of America today, John Stossel speaks out.

http://townhall.com/columnists/JohnStossel

http://video.google.com/videoplay?docid=1876894381231272307#

Endnotes

1   Survey by the MIT Laboratory for Energy and the Environment (LFEE)

http://sequestration.mit.edu/pdf/LFEE_2005-001_WP.pdf
http://sequestration.mit.edu/pdf/Tom_Curry_Thesis_June2004 Entries are removed after a few years so may not be available.

2   Who Killed the Electric vehicle?

http://www.ocregister.com/ocregister/entertainment/movies/newreleases/article_1197832.php

3   *Scientific American*, May, 2004 issue

Section VI

Appendix

This section contains significant information about and from people, organizations and manufacturers who may affect the development and outcome of any solutions to our changing energy economy. It also contains descriptions of new products and processes developed by them. Most of the information is available on the Internet so links to the various web sites are included.

Much information is available that changes almost daily so Internet searches on almost any of the subjects will turn up new and sometimes conflicting information.

Hopefully some of these will help us escape the petroleum dragon now breathing fire down our economic and environmental necks.

American imagination, ingenuity and determination are still alive and well.

What follows are some examples of what private (and sometimes very small) American companies and industries are doing right now to develop the technologies needed for the solutions described in this book. This information is taken from the Internet and is usually written by the people involved in the particular entrepreneurial effort and products described. The author of this book has not researched any of these claims or descriptions and therefore has only a perfunctory opinion about their correctness or practicability. Therefore, their inclusion herein does not indicate any kind of approval or endorsement by the author. Many magic, energy out of nothing, and secret smoke and mirror articles can be found if one looks even a little. Only those articles that make rational sense to the author were included so read accordingly. The author asked for and received permission to include many of these articles and has done so, usually wholly.

Excerpts were included from those articles from sources that did not respond and are so noted. A few made it so difficult to obtain permission (GM and Wikipedia) or so expensive (*New York Times* of course) that they also were only excerpted. News releases were used in their entirety. The link to each article is listed at the start of the inclusion, right beneath the dividing line.

Several major manufacturers were also contacted several times and asked for information that could be included in the book. Neither Ford nor Chrysler ever replied. GM responded, but declined to provide anything other than advertising brochures. Toyota, Nissan, Mazda, and Hyundai each responded but declined providing any information saying new product development was kept under secrecy. I guess all the auto manufacturers are holding their cards tightly against their chests. Many small manufacturers also referred me to advertising or the Internet and information already in my possession. There were but two exceptions. Capstone Turbine Corporation of California provided considerable information and specification sheets on their micro turbines. Dr. Jerry M. Woodall and writer Emil Venere of Purdue University also provided much information about their intriguing process for generating hydrogen on the spot. Apparently most companies do not think much of books for publicity. I wonder if their attitudes will change should this become a big seller?

Included in this appendix are some of the author's commentaries. These comments are identified and in italics.

The first article covers President Bush's remarks about energy, fuel and vehicles from his 2006 State of the Union Address.

http://www.whitehouse.gov/news/releases/2006/01/20060131-6.html

News Release by the Press Secretary

In His State of the Union Address, President Bush Outlined the Advanced Energy Initiative to Help Break America's Dependence on Foreign Sources of Energy. The President has set a national goal of replacing more than 75% of our oil imports from the Middle East by 2025. With America on the verge of breakthroughs in advanced energy technologies, the best way to break the addiction to foreign oil is through new technology. Since 2001, we have spent nearly $10 billion to develop cleaner, cheaper, and more reliable alternative energy sources. Tonight, the President announced the Advanced Energy Initiative, which provides for a 22% increase in clean energy research at the Department of Energy (DOE). The Initiative will accelerate our breakthroughs in two vital areas; how we power our homes and businesses; and how we power our automobiles.

Changing the Way We Power Our Homes and Businesses

The Administration Will Work to Diversify Energy Sources for American Homes and Businesses. Accelerating research in clean coal technologies, clean and safe nuclear energy, and revolutionary solar and wind technologies will reduce overall demand for natural gas and lead to lower energy costs. The President's Advanced Energy Initiative proposes speeding up research in three promising areas:

The President's Coal Research Initiative: Coal provides more than half the Nation's electricity supply, and America has enough coal to last more than 200 years. As part of the National Energy Policy, the President committed two billion dollars over ten years to speed up research in the use of clean coal technologies to generate electricity while meeting environmental regulations at low cost. To tap the potential of America's enormous coal reserves, the President's 2007 Budget includes $281 million for development of clean coal technologies, nearly completing the President's commitment four years ahead of schedule.

In the author's opinion, sequestering carbon dioxide, and in particular carbon dioxide from the burning of coal in power plants, will be a monumental and expensive task. The results of developing and implementing such technology will hardly be worth the costs if indeed the problem is ever solved. I believe investing in other proven geothermal or nuclear technology or even the new wave motion systems will cost far less and get better results than trying to solve the so far unsolvable problem of sequestering carbon dioxide.

The President's 2007 Budget Includes $54 Million for the FutureGen Initiative: The FutureGen initiative is a partnership between government and the private sector to develop innovative technologies for an emissions free coal plant that captures the carbon dioxide it produces and stores it in deep geologic formations.

The President's Solar America Initiative: The 2007 Budget will propose a new $148 million Solar America Initiative—an increase of $65 million over fiscal year 2006—to accelerate the development of semiconductor materials that convert sunlight directly to electricity. These solar photovoltaic or *PV* cells can be used to deliver energy services to rural areas and can be incorporated directly into building materials, so that future *zero energy* homes can be built that produce more energy than they consume.

Expanding Clean Energy from Wind: The 2007 Budget includes $44 million for wind energy research—a $5 million increase over FY06 levels. This will help improve the efficiency and lower the costs of new wind technologies for use in low-speed wind environments. Combined with ongoing efforts to expand access to Federal lands for wind energy development, this new funding will help dramatically increase the use of wind energy in the United States.

Changing the Way We Power Our Automobiles: We Are on the verge of dramatic improvements in how we power our automobiles, and the president's initiative will bring those improvements to the forefront.

The United States must move beyond a petroleum-based economy and develop new ways to power automobiles. The President wants to accelerate the development of domestic, renewable alternatives to gasoline and diesel fuels. The Administration will accelerate research in cutting-edge methods of producing *cellulosic ethanol* with the goal of making the use of such ethanol practical and competitive within 6 years. The Administration will also step up the Nation's research in better batteries for use in hybrid and electric cars and in pollution-free cars that run on hydrogen.

In the author's opinion, ethanol will not be any more than a stopgap fuel while we convert to using more viable and less costly systems. It is also doubtful the hydrogen fuel cell vehicle will ever be a practical reality for the many reasons described in the main section of this book.

The Biorefinery Initiative: To achieve greater use of *homegrown* renewable fuels in the United States, advanced technologies need to be perfected to make fuel ethanol from cellulose (plant fiber) biomass, which is now discarded as waste. The President's 2007 Budget will include $150 million—a $59 million increase over fiscal year 2006—to help develop biologically-based transportation fuels from agricultural waste products, such as wood chips, corn stalks, or switch grass. Research scientists say that accelerating research into *cellulosic ethanol* can make it cost-competitive by 2012, offering the potential to displace up to 30 percent of the Nation's current fuel use.

Developing More Efficient Vehicles: Current hybrids on the road run on a battery developed at the DOE. The President's plan would accelerate research in the next generation of battery technology for hybrid vehicles and *plugin hybrids*. Current hybrids can only use the gasoline engine to charge the onboard battery. A *plugin hybrid* can run either on electricity or on gasoline, and can be plugged into the wall at night to recharge its batteries. These vehicles will enable drivers to meet most of their urban commuting needs with virtually no gasoline use. Advanced battery technologies offer the potential to significantly reduce oil consumption in the near-term. The 2007 Budget includes $30 million—a $6.7 million increase over FY06—to speed up the development of this battery technology and extend the range of these vehicles.

Author's note: This is one area where I believe research money could be well spent although it may be too little and too late. Several talented entrepreneurs have developed radical new battery technology that is already hitting the market without government help (or control). Good old American ingenuity and drive are at it again. Some of these batteries are already being installed in all-electric vehicles coming to market as this is written. Information about these batteries is included in several articles in this section.

The Hydrogen Fuel Initiative: In his 2003 State of the Union address, President Bush announced a $1.2 billion Hydrogen Fuel Initiative to develop technology for commercially viable hydrogen-powered fuel cells, which would power cars, trucks, homes, and businesses with no pollution or carbon dioxide. Through private-sector partnerships, the Initiative and related FreedomCAR programs will make it practical and cost-effective for Americans to use clean, hydrogen fuel cell vehicles by 2020. The President's 2007 Budget will provide $289 million—an increase of $53 million over FY06—to accelerate the development of hydrogen fuel cells and affordable hydrogen-powered cars. Through the President's program, the cost of a hydrogen fuel cell has been cut by more than 50 percent in just four years.

In the author's opinion, the hydrogen fuel cell vehicle will never be a practical, cost-effective reality. In spite of the several radical new technologies described in these pages, the overall costs promise to be much higher than that of several other technologies including micro turbine-powered PHEVs and new battery technologies powering pure-electric vehicles or EVs.

America must Act Now to Reduce Dependence on Foreign Sources of Energy. There are an estimated 250 million vehicles on America's highways, and Americans will purchase more than 17 million vehicles this year. It will take approximately 15 years to switch America's automobiles over to more fuel efficient technologies. The sooner breakthroughs are achieved, the better for America.

In the author's opinion: If American entrepreneurs are free to use the systems described in this book without too much interference from government and big business, they can accomplish nearly complete energy self-sufficiency in ten years. This will not only provide a tremendous boost to our economy, but can also stem the growing hemorrhage of billions of dollars now going out of our economy for imported oil.

The President's Advanced Energy Initiative Will Build on the Progress Made Since 2001. The administration has worked to ensure affordable, reliable, secure, and clean sources of energy. In 2001, the President put forward his National Energy Policy, which included over 100 recommendations to increase domestic energy supplies, encourage efficiency and conservation, invest in energy-related infrastructure, and develop alternative and renewable sources of energy. Over the past four years, the Administration has worked to implement these recommendations and improve the Nation's energy outlook.

Last Summer, the President Signed the First Comprehensive Energy Legislation in over a Decade. The Energy Policy Act of 2005 is strengthening America's electrical infrastructure, reducing the country's dependence on foreign sources of energy, increasing conservation, and expanding the use of clean renewable energy.

In the author's opinion, America's independent, creative genius could solve this in short order if the government would offer rewards to companies, individuals, and institutions for providing practical solutions. The government billions now funding research (often in pork barrel grants and funds) could much better be spent on rewards for results. It is axiomatic that people will do that which promises the biggest rewards for them. Offer rewards for results and results will be forthcoming. Offer rewards for effort and effort will be forthcoming and without necessarily providing results. In fact, some research projects funded by grants deliberately avoid producing results or solving their assigned problem knowing that will end the project and often their jobs and sources of income. One of the biggest rewards and greatest incentives to rapid success would be a <u>no tax</u> period for companies that develop and produce real, practical, workable solutions for nonpetroleum energy. Zero income and zero capital gains taxes for these companies and individuals for ten years of profitability would be a potent incentive. Since this would only come into effect when and if the company became profitable it would be a more powerful incentive than any research grant.

NOTE: The new administration has removed all Internet available records from the previous administration so none of this is accessible any longer. How do you like them apples? I wonder what other information they do not like has been removed and destroyed? Remember Sandy Burger's notorious removal of documents? The media certainly buried that in a hurry. If a conservative or Republican had done such an illegal thing it would still be getting front page play. Don't you just love that impartial main-stream media?

web.mit.edu/newsoffice/2003/hydrogen-0305.html

Hydrogen vehicle won't be viable soon, study says

Source of excerpts—Malcolm A. Weiss and John B. Heywood of MIT's Laboratory for 21st-Century Energy—News Release about a study of the hydrogen fuel cell vehicle recently released by MIT's Laboratory for Energy and the Environment (LFEE)—Boston, March 11, 2003

Big Oil can sleep easy for another decade or two should we wait for hydrogen fuel cells to wean the West off its addiction to oil.

Hydrogen Vehicle Won't Be Viable Soon, Study Says—by Nancy Stauffer, spokesperson for Laboratory for Energy and the Environment at MIT. "Even with aggressive research, the hydrogen fuel cell vehicle will not be better than the diesel hybrid (a vehicle powered by a conventional engine supplemented by an electric motor) in terms of total energy use and carbon dioxide emissions by 2020," says a study recently released by the MIT Laboratory for Energy and the Environment (LFEE).

Hybrid and PHEV vehicles are already appearing on the roads mainly because they use existing technologies and infrastructure. Adoption of the hydrogen-based vehicle will require major infrastructure changes to make compressed hydrogen available. If adopted, it will also make obsolete all our existing vehicles and small engines. If we need to curb carbon dioxide within the next twenty years, improving mainstream gasoline and diesel engines and transmissions and expanding the use of hybrids and PHEVs is one way to go.

The results from a systematic and comprehensive assessment of a variety of engine and fuel technologies as they are likely to be in 2020 after intense research but no real breakthroughs, show little hope hydrogen will ever be applicable as a fuel for transportation. Other technologies are already proven that can do a better job in this area. They are not only available, but far less expensive especially considering the need for the required new infrastructure.

Many other types of hybrid and tribrid vehicles or PHEVs and all-electric battery-powered EVs are available. These are much better overall than the hydrogen fuel cell vehicle using virtually any criteria one could develop. Many of these are described in the following pages. In fact, one such vehicle has already been shown in prototype form by GM, at the Detroit motor show, the Volt concept car described in GM's press release.

The study was released just a month after the Bush administration announced a billion-dollar initiative to develop commercially viable hydrogen fuel cells and a year after establishment of the government-industry program to develop the hydrogen fuel cell-powered Freedom Car.

This assessment is on top of a study done in 2000, which likewise concluded that the much-touted hydrogen fuel cell system had many drawbacks and might never be practical. This time, the MIT researchers even used optimistic fuel cell-performance assumptions cited by some fuel cell advocates, and the conclusion remained the same.

The hydrogen fuel cell vehicle may have low emissions and energy use on the road—but converting a hydrocarbon fuel such as natural gas or gasoline into hydrogen to fuel this vehicle uses substantial energy and emits even more carbon dioxide.

"Ignoring the emissions and energy use involved in making and delivering the fuel and manufacturing the vehicle gives a misleading impression," said Weiss.

Still, these researchers do not recommend stopping work on the hydrogen fuel cell. "If auto systems with significantly lower carbon dioxide emissions are required in, say, 30 to 50 years, hydrogen is the only major fuel option identified to date," said Heywood.

The author says this is just not true. Many organic, renewable fuels producing zero-net carbon dioxide and few pollutants and many applicable systems are available to produce, distribute and use them with existing technologies. Improvements and innovative development of battery technology could even overshadow renewable fuels as portable energy. Hydrogen is certainly not the only major fuel option available for autos. Other fuel systems are not only available, but are cheaper, far easier to convert, and produce almost no pollution or carbon dioxide if properly designed. If sequestering carbon emissions was at all practical, doing so for existing vehicles and power users would instantly solve the carbon dioxide problem.

The study says:

The hydrogen must, of course, be produced without making carbon dioxide emissions, hence from a noncarbon source such as solar energy or from conventional fuels while sequestering the carbon emissions. The assessment highlights the advantages of the hybrid, a highly efficient approach that combines an engine (or a fuel cell) with a battery and an electric motor.

Continuing to work on today's gasoline engine and its fuel will bring major improvements by 2020, cutting energy use and emissions by a third compared with today's vehicles. But aggressive research on a hybrid with a diesel engine could yield a 2020 vehicle that is twice as efficient and half as polluting as that evolved technology, and future gasoline engine hybrids will not be far behind.

This is but one of many well-considered articles describing the inadequacies, flaws and dangers associated with the development of a viable hydrogen fuel cell power system. No one has yet tackled the most formidable problem in the hydrogen economy. One that has hardly even been mentioned is the infrastructure required to generate, store and deliver the hydrogen to vehicles safely and economically. That will not happen in the foreseeable future because the cost of the infrastructure is prohibitive.

This book proposes several alternative systems that are better in almost every way than the hydrogen fuel cell and does not require the monstrous investment in development and infrastructure. These systems can be merged into those presently in use with minimal displacement of existing vehicles, infrastructure, manufacturing processes or public understanding.

---

http://today.uci.edu/news/release_detail.asp?key=694

Excerpt—Go to this site to read the entire article.

National Fuel-cell Research Center at U C Irvine to test new, efficient and clean power generation technology

Irvine, Calif., June 15, 1999

Edison Technology Solutions has chosen the National Fuel-cell Research Center at UC Irvine to test a new, highly efficient type of power plant. Under construction at Siemens Westinghouse Power Corp., the *hybrid* system integrates a fuel-cell with a gas micro turbine and is expected to generate electricity at efficiencies believed unreachable in small-sized plants just a few years ago.

If successful, the technology will produce electricity at low cost. Urban smog-forming pollutant emissions from power plants could be reduced to undetectable levels and the emission of carbon dioxide from those plants could fall by more than half as well, according to Scott Samuelsen, director of the National Fuel-cell Research Center.

Researchers from UCI's National Fuel-cell Research Center and engineers from Edison Technology Solutions, part of Edison International, are preparing the test site and plan for the 250-kilowatt plant, with full operation scheduled for October.

Author's note: It seems that Technology Solutions has closed their operations and is no longer in business. A quick search of the Internet found no explanation for this or where the assets of ETS went. I note this only because this appeared to be a promising technology that has seemingly disappeared. A thorough search of the Internet revealed no further information about this.

Their Web site: **http://www.edisontec.com/** is no longer active. *A bit of research uncovers the following: The registrant is Jill Yang - 7F, No. 55, Ln 356, Long-Jiang Rd., Taipei 104, Taiwan. It was registered on 19-Oct-07. It expires on 19-Oct-09, and was last Updated on: 11-Jul-08 through: GoDaddy.com, Inc.* See instead:

www.cd3wd.com/cd3wd_40/JF/JF_OTHER/.../REILLY.PDF

http://fuelcellbus.georgetown.edu/files/MethanolFromCoalFinalRepor t04-2004.pdf

An Investigation of the Feasibility of Coal-based Methanol for Application in Transportation Fuel-cell Systems

Used with permission

Submitted to Georgetown University

Prepared by: Vernon Roan, Principal Investigator, Daniel Betts, Amy Twining, Khiem Dinh, Paul Wassink and Timothy Simmons

University of Florida, Gainesville, Florida April 2004

ABSTRACT

This University of Florida report examines some of the issues relating to the future of transportation fuels. The study gave particular emphasis to the likely paths that could potentially provide transition from current petroleum-based liquid transportation fuels to a full-blown *hydrogen economy*. Paramount among the potential transition path questions are whether an intermediate fuel will be necessary to ease the petroleum to hydrogen economy transition and, if so, what are likely candidate fuels?

It was obvious from the outset that the quantities of energy associated with transportation are already enormous globally and domestically, and that these energy requirements are likely to increase, probably dramatically, as China and other large-population countries expand their automobile and truck use to ever larger fractions of the population.

These transportation energy requirements eliminate all but a few options for fuel feedstocks and provide indication of the magnitude of cost and difficulty in achieving a true hydrogen-transportation fuel capability.

Author's special note: *A <u>hydrogen economy</u> for fuel-cell vehicles is an unlikely eventuality for several reasons covered in other sections of this book. As Matthew Wald writes in Scientific American, "Only in an economy where quantities are so small that price is no factor will the hydrogen fuel-cell be economically feasible."*

Based on the limited options for transportation fuel feedstocks, consideration in this study was limited to natural gas, coal and water. Natural gas is already being used, in a limited way, directly as a transportation fuel, so it is obvious that natural gas is a potential intermediate fuel. Natural gas is also the primary feedstock, at present, for the production of both hydrogen and methanol, which implies that continuing and expanding this role, is a consideration.

Coal is used as a fuel, primarily in electrical power generation, accounting for slightly more than half of all power generation in the United States. It is also far more abundant, energy wise, in the United States than either petroleum or natural gas. Even though coal is not widely used as a transportation fuel feedstock, commercial processes have been developed to produce both methanol and hydrogen from coal. This is a scenario of interest since these processes also carry the possibility of transportation energy independence.

Water is obviously a virtually unlimited feedstock for the production of hydrogen. Unfortunately, producing hydrogen in this manner requires more electrical energy input in hydrogen production than can be reclaimed through the utilization of the chemical energy available in hydrogen as a fuel. In other words, there is a net loss in available energy when hydrogen is obtained from water. Therefore, an inexpensive and bountiful source of electricity would be required to harvest hydrogen from water in the magnitudes required to fulfill the expected requirements of the *hydrogen economy*. The study makes first order comparisons of these fuel options by considering the major costs associated with production and transportation of the various fuels. This study does not include a complete economic analysis but is based on what are believed to be ranges of reasonable assumptions for the time periods considered. Projections are also made based on historical data of fuel taxes, which will be necessary irrespective of fuel utilization.

The results of the study, while not surprising, point out some important considerations and likely future events. Below is a list of some of the study's conclusions.

1. The amount of petroleum imported into the United States is larger than during the energy crisis of the 1970s and is increasing. Reducing this dependency on foreign petroleum is in the best long-term interests of the U.S.

2. Natural gas and coal are the only energy sources currently available in quantities comparable to petroleum for transportation. The cost of natural gas is likely to increase if demand increases as expected. Also, more natural gas will have to be imported into the U.S. to meet this increased demand. Thus natural gas may not be an appropriate feedstock for future alternative fuels if the goal is to reduce dependence on foreign energy sources.

3. Recoverable reserves of coal will last at least five times as long as technically recoverable natural gas or petroleum in the U.S.

4. Methanol is the most desirable liquid hydrocarbon fuel for fuel-cells and can be effectively utilized in internal combustion engines using existing technologies. While all alternative fuels are expected to be more expensive to the consumer than present-day gasoline, methanol produced from coal is likely to be the least expensive of the fuels considered, if natural gas prices increase as projected.

Author's note: The article's proposed process for making methanol from hydrogen and carbon dioxide described elsewhere in this book could be far cheaper than producing it from coal and without adding carbon dioxide to the atmosphere. This process could actually remove carbon dioxide from the atmosphere.

Future events are impossible to predict with a reasonable degree of certainty. Nevertheless, the results of this study show that the likely time between the dramatic decline in easily recoverable petroleum and significant movement into a hydrogen economy could leave a gap of decades. It also appears that this gap cannot be filled by inexpensive natural gas, especially from domestic production. If these projections are correct, then it appears that methanol produced from coal could be the most attractive option from the standpoints of cost and domestic availability. Environmental and safety issues are clearly associated both with coal as a feedstock and methanol as a widely distributed transportation fuel that would have to be satisfactorily resolved. However, if such issues can be resolved, methanol produced from coal could utilize much of the existing fuel infrastructure. Also, long-range fuel price stability may be expected due to the price stability of coal. Methanol is also the most desirable hydrocarbon fuel for fuel-cell vehicles as well as a proven fuel for internal combustion engines.

Finally, depending on the assumptions made, the study suggests that the cost of coal-based methanol per gallon of gasoline energy equivalent could be much less than the cost of hydrogen or even natural gas. This could be very important since all of the other scenarios suggest a fuel cost much higher.

Another chapter from the methanol from coal report of the U. of Florida:

CHAPTER 1—INTRODUCTION

Reducing dependency on foreign supplies of petroleum is in the best long-term interests of the United States. Since transportation systems consume the majority of petroleum, obvious ways to move in this direction are to develop more energy-efficient transportation systems and/or to utilize more nonpetroleum-based alternative fuels. Transportation systems can be made more energy-efficient by reducing the energy demands of the vehicle (e.g., lighter materials, better aerodynamics, etc.) or improving the energy conversion efficiency of the power source (e.g., improved internal combustion engines, hybrid systems, fuel-cells, etc.). These approaches are all being vigorously pursued by vehicle manufacturers as well as many other researchers and United States Government agencies, especially the Department of Energy.

Of the alternative fuels being considered, many people view hydrogen as the ultimate long-term alternative fuel. However, while hydrogen has some desirable attributes, many difficult issues must be resolved. That might make the time scale for a *hydrogen economy* far longer than the time scale of readily available and affordable petroleum. In addition, fuels that can be stored as liquids offer major advantages for transportation applications. Of the alternative liquid fuels, in many ways the most promising is methanol.

Most methanol production (and hydrogen production) currently uses natural gas as the feedstock. Natural gas is a versatile feedstock since it can be used directly as a fuel, including transportation fuel, and is a feedstock for producing methanol and hydrogen. Most home heating, many chemical productions, and much power generation depend on inexpensive natural gas. However, even without large-scale use as a transportation fuel, the demand for natural gas has been steadily increasing while domestic production has been nearly flat in recent years.

Right now significant quantities of stranded offshore natural gas are being either *flared* or underutilized. In principle, this natural gas could be used in the United States by various means, including:

1. liquefying and shipping to the United States as liquid natural gas (LNG).

2. using this gas as a feedstock for producing hydrogen that could then be shipped to the United States.

3. using this gas as a feedstock for producing methanol that could be shipped to the United States as a liquid.

Of these three alternatives, the offshore production of methanol offers the most favorable economics and overall best feasibility for a transportation fuel. While natural gas is the preferred feedstock for the most economical production of hydrogen and methanol, almost any plentiful hydrocarbon can be used for this purpose.

Probably the best alternative in this group is coal. Pilot plants have been built, and are operational to produce methanol from coal. Both liquid phase and gaseous phase processes have been successfully demonstrated for such coal-based methanol production. While such production might be somewhat more costly than current stranded natural gas methanol production, it could offer a viable long-term alternative using plentiful domestic coal supplies as feedstocks. This is true especially if natural gas demand (and price) continues to increase as expected.

Many studies have already been completed which demonstrate that there is no *ideal* fuel to replace petroleum-based fuels for transportation systems. However, it has also been demonstrated both through studies and experimental fuel-cell vehicles that methanol has many desirable characteristics. Indeed, excluding pure hydrogen, it is probably the most fuel-cell friendly fuel, and excluding petroleum-based fuels, it is probably the liquid fuel with the best outlook for mass production and infrastructure possibilities. It is also readily adaptable to virtually any type of heat engine including spark ignition, compression ignition (Diesel) and gas micro turbine engines.

This study was undertaken to decide the feasibility of using methanol for transportation systems for near-term (next 10 years), midterm (10 to 25 years), and longer term (more than 25 years). Specifically, the economics of methanol versus hydrogen as alternatives to petroleum-based transportation fuels, and the likely feedstocks for the alternative fuels through these time periods were considered. The prior supposition was that the offshore stranded natural gas methanol production would probably provide the lowest cost option for near term (and perhaps longer) methanol. If natural gas supplies dwindled and/or if big increases in demand for natural gas continued, the costs would increase accordingly. Coal-based methanol would presumably become economically practical. Of course, environmental issues are involved with the production of methanol (or hydrogen) from coal that would have to be resolved prior to large-scale production. For example, more carbon dioxide would be generated than from natural gas-based methanol. In addition, sulfur, particulate matter, toxic compounds, and solid residue could cause problems. On the positive side, major social, political, and security advantages will result from using domestic energy sources, not to mention the huge economic boon.

Efforts were made to include realistic economic considerations in this study. These factors are often overlooked or neglected. For example, it was assumed that it will be necessary for projected state and federal tax revenues to be compatible with funding necessary for continued highway construction, repair and other current applications of the tax dollars.

Although it was beyond the scope of this study to perform detailed economic analyses, efforts were made to account for differences in costs due to alternative methods for producing and transporting various fuels. For example, the relative costs of producing methanol near coal mines and transporting the methanol versus transporting coal to distributed methanol production sites.

It was not a prior assumption nor is it a conclusion of this study that methanol is the fuel of choice to replace petroleum-based fuels. There are environmental, safety, economic, and other issues which must be satisfactorily resolved prior to consideration of mass production and distribution of methanol. In addition, there are other fundamental issues that must be addressed.

It is important to evaluate whether methanol or any interim fuel is feasible for widespread use en route to a *hydrogen economy*. That is, could the economy support two major infrastructure changes during such a transitional time period? If so, what would the time scale have to be for this to be economically feasible? Clearly, the implication is that the longer the time lapse prior to a self-sustaining hydrogen economy, the more feasible, indeed more likely necessary, it will be to develop an interim alternative fuel and supporting infrastructure. It should also be recognized that a liquid alternative fuel might continue to be needed even as the hydrogen economy evolves.

There are many questions not addressed by this study. The study objectives were centered on comparisons of the feasibility of methanol, especially coal based methanol, versus other alternative fuels for the time periods considered as alternatives to petroleum-based fuels.

Author's special note: *Those who extol the virtues of coal either ignore the large quantities of carbon dioxide the use of coal would add to the atmosphere or think that is not a problem. Others say, "All we need is an economical method to sequester carbon dioxide." So far, development and implementation of such a method on a practical, commercial scale have eluded all efforts. Nature's own green plants are still the only known practical method of sequestering carbon dioxide.*

http://www.autobloggreen.com Search on GM Volt

Is GM's Volt Concept Car the first PHEV and is it about to become a reality?

PRESS RELEASE

NOTE: After this release, GM announced production of the Volt with an introduction date of 2010. Considering the fact that the government and UAW are now running GM, any successful future activities are doubtful. I see this takeover action as the death knell for GM and Chrysler. Any new cars they produce will doubtless require massive amounts of federal subsidies. I doubt GM or Chrysler will ever again be viable, profitable enterprises until and unless they become truly private enterprises once more. All references to GM in this book should now be read with this in mind.

For several months now rumors have been rampant about an electric vehicle that General Motors would unveil at the Detroit Auto Show. That vehicle is now real, in the form of the Chevrolet Volt. The Volt is the first vehicle application of the GM's new E-Flex platform. Volt is a C-Class sized four door sedan roughly the size of a Cobalt.

In spite of the presence of an internal combustion engine, GM does not call this vehicle a hybrid. In fact, they consider it an EV with range extending capability. The engine is a turbo charged, 1.0L three cylinder engine with 71 hp that has no mechanical connection to the wheels. The ICE runs at about 1800 rpm and drives a 53 kW generator that charges the lithium-ion battery pack. The engine starts and stops automatically as needed to charge the battery.

The battery pack provides power to a 161 hp (120 kW) electric motor that's connected to the front wheels to provide the motive force. It's the same motor that's used in the fuel-cell Equinox. The Lithium ion battery has a peak output of 136 kW and a total capacity of 16 kWh. The battery can be charged by plugging it in to any standard 110 V outlet and is fully charged in about 6-6.5 hours. There are two plugs, one on each side of the car, to facilitate home charging.

The Volt has a range of about 40 miles on the battery alone which might not seem like much. But, considering that most people drive fewer miles than that per day, it should mean that a lot of drivers will never use a drop of gas on their daily commute. However, when the fuel tank is filled to its capacity of 12 U.S. gallons of gas, the Volt has a range of 640 miles. In addition, the Volt ICE is fully flex-fuel capable and can run on any combination of gasoline or ethanol up to E85. The power-train is sized to achieve 0-60 mph acceleration of about 8.5 seconds.

For a customer driving about 40 miles a day or about 15,000 miles a year, compared with a 30 mpg car, the Volt would save about 500 gallons of gasoline per year. If the car is charged every night, the driver should be able to achieve that mileage using virtually no gasoline. That same example would also save 4.4 metric tons of carbon dioxide every year from each car. Another example of a driver commuting 60 miles a day would achieve an equivalent mileage of 150 mpg based on the engine running for the last 20 miles in a charge sustaining mode. As the driver's mileage drops down toward that 40-mile threshold, the equivalent mileage rises toward infinity. The ICE/generator combo has enough power to keep the car going when cruising at 70 mph and after the 30 minutes of running, the battery will be topped up.

GM's goal was to create an electric car that would not force users to plan their travel around the next charging session, while still providing all the capabilities of a standard four-door, standard compact car and produce it in quantities of 100,000+ per year. They seem to have succeeded at the first part of this. Now the big question is when can we buy one? Here things get decidedly murky.

In GM's development process, a program is not considered a real production intent vehicle until a vehicle line executive is assigned. The Volt has a VLE in the person of Tony Posawatz, so it is intended for showrooms, not just the show circuit. The only thing that is not quite real at this point is the timing. The hold-up is that darned battery. At this point no car-maker in the world has yet publicly committed to building a car powered by Li-ion batteries in any significant quantities (Tesla is now delivering production roadsters, has announced plans to introduce a sedan and possibly other more main-stream models, and has just announced a joint venture with Toyota to participate in the resurrection of the RAV4 EV by 2012). Regardless of the claims of battery makers, the technology to build an affordable battery that will last 100,000 miles, with minimal degradation of performance has yet to be demonstrated. GM is looking at a number of potential suppliers, but so far hasn't committed to any. No pricing is available at this point, but the base price is almost certain to be more than a comparable Cobalt or Focus. However, they want to price it so that total operating cost of the vehicle and fuel costs are comparable or less than current cars. Given, the efficiency of such a vehicle, it should allow quite a bit of latitude, as long as customers buy into that concept.

GM made sure to emphasize that the Volt and E-Flex are not science fair projects or PR stunts. For the sake of GM and the domestic industry as a whole, they better bring something like this to market sooner rather than later.

Read all about the E-Flex system:

http://www.autobloggreen.com

Search on e-flex-platform, Chevy Volt, and GM EV1.

See a comparison of the Volt to the EV1:

The following is the rest of GM's press release on the Volt:

The Chevrolet Volt concept sedan, powered by the E-flex System—GM's next-generation electric propulsion system—and sporting an aggressive, athletic design, could nearly eliminate trips to the gas station.

The Chevrolet Volt is a battery-powered, four-passenger electric vehicle that uses a gas engine to create additional electricity to extend its range. The Volt draws from GM's previous experience in starting the modern electric vehicle market when it launched the EV1 in 1996, according to GM Vice Chairman Robert A. Lutz:

"The EV1 was the benchmark in battery technology and was a tremendous achievement," Lutz said. "Even so, electric vehicles, in general, had limitations. They had limited range, limited room for passengers or luggage, couldn't climb a hill or run the air conditioning without depleting the battery, and had no device to get you home when the battery's charge ran low.

"The Chevrolet Volt is a new type of electric vehicle. It addresses the range problem and has room for passengers and their stuff. You can climb a hill or turn on the air conditioning and not worry about it. The Volt can be fully charged by plugging it into a 110-volt outlet for approximately six hours a day. When the lithium-ion battery is fully charged, the Volt can deliver more than 60 city kilometers of pure electric vehicle range."

When the battery is depleted, a 1.0-liter, three-cylinder turbo charged engine spins at a constant speed, or revolutions per minute (rpm), to create electricity and replenish the battery. According to Lutz, this increases the fuel economy and range.

"If you lived within 50 km from work (one hundred km round trip) and charged your vehicle every night when you came home or during the day at work, you would get fuel consumption of 1.6 liters per one hundred km," Lutz said. "More than half of all Americans live within around 30 km of where they work (60 km round trip). In that case, you might never burn a drop of gas during the life of the car." In the event a driver forgets to charge the vehicle or goes on a vacation far away, the Volt would still get 50 mpg by using the engine to convert gasoline into electricity and extending its range up to 650 miles, more than double that of today's conventional vehicles. In addition, the Chevrolet Volt is designed to run on E85, a fuel blend of 85 percent ethanol and 15 percent gasoline.

A technological breakthrough required to make this concept a reality is a large lithium-ion, or other new technology battery. This type of electric car, which the technical community calls an *EV range-extender*, would require a battery pack that weighs nearly 400 pounds (181 kg). Some experts predict that such a battery, or a similar battery, could be production-ready by 2010 to 2012.

Jon Lauckner, GM vice president of Global Program Management, said the Volt is uniquely built to accommodate a number of advanced technology propulsion solutions that can give GM a competitive advantage.

"Today's vehicles were designed around mechanical propulsion systems that use petroleum as their primary source of fuel." Lauckner said. "Tomorrow's vehicles need to be developed around a new propulsion architecture with electricity in mind. The Volt is the first vehicle designed around GM's E-flex System.

"That's why we are also showing a variant of the Chevrolet Volt with a hydrogen-powered fuel-cell, instead of a gasoline engine EV range-extender. Or, you might have a diesel engine driving the generator to create electricity, using biodiesel. Finally, an engine using 100 percent ethanol might be factored into the mix. The point is, all of these alternatives are possible with the E-Flex System."

Author's note: *Mr. Lauckner is either ignorant or deliberately ignoring the fact that there is no such thing as 100% ethanol other than under very difficult manufacturing and expensive storage systems. Pure ethanol has such a powerful affinity for water it will draw it out of the air until it becomes an azeotropic mixture of 95% ethanol and 5% water. That's why E85 flex-fuel is 85% ethanol and 15% gasoline.*

"The Volt concept car is built on a modified future architecture," Lauckner said, "similar to the one GM uses for current small cars, such as the Chevrolet Cobalt and HHR".

According to Larry Burns, GM vice president for research and development and strategic planning, the world's growing demand for energy and its dependence on oil for transportation is the common theme behind today's headlines:

"Whether your concern is energy security, global climate change, natural disasters, the high price of gas, the volatile pricing of a barrel of oil and the effect that unpredictability has on Wall Street—all of these issues point to a need for energy diversity," said Burns. "Today, there are more than 800 million cars and trucks in the world. In 15 years, that will grow to 1.1 billion vehicles. We can't continue to be 98 percent dependent on oil to meet our transportation needs. Something has to give. We think the Chevrolet Volt helps bring about the diversity that is needed. If electricity met only 10 percent of the world's transportation needs, the impact would be huge."

GM's E-flex system moves automobile toward new electric age

GM's E-flex System enables multiple propulsion systems to fit into a common chassis, using electric drive to help the world diversify energy sources and establish electricity from the grid as one of those sources.

"The DNA of the automobile has not changed in more than 100 years," said Burns. "Vehicles still operate in pretty much the same fashion as when Karl Benz introduced the horseless carriage' in 1886.

"While mechanical propulsion will be with us for many decades to come, GM sees a market for various forms of electric vehicles, including fuel-cells and electric vehicles using gas and diesel engines to extend the range. With our new E-flex concept, we can produce electricity from gasoline, ethanol, biodiesel or hydrogen.

"We can tailor the propulsion to meet the specific needs and infrastructure of a given market. For example, somebody in Brazil might use 100 percent ethanol (E100) to power an engine generator and battery. (*E100 is an impossible fuel - see author's note earlier on this page*) A customer in Shanghai might get hydrogen from the sun and create electricity in a fuel cell. Meanwhile, a customer in Sweden might use wood to create biodiesel."

The Chevrolet Volt is just the first variant of the E-flex System. The Volt uses a large battery and a small, 1.0-liter turbo charged gasoline engine to produce enough electricity to go up to 1030 km and provide triple-digit fuel economy. GM will show other variations of the propulsion systems at future auto shows.

"GM is building a fuel-cell variant that mirrors the propulsion system in the Chevrolet Sequel (fuel-cell concept)," Burns said. "Instead of a big battery and a small engine generator used in the Volt, we would use a fuel-cell propulsion system with a small battery to capture energy when the vehicle brakes. Because the Volt is so small and lightweight, we would need only about half of the hydrogen storage as the Sequel to get around 480 km of range." Future concepts might incorporate diesel generators, biodiesel and other RN fuels.

Author's note: *Micro turbine generators might be an even better alternative, especially considering they could run on virtually any liquid or gaseous fuel. Environmentally conscious vehicles can be aesthetically appealing.*

With exterior proportions associated more with classic sports cars, the Chevrolet Volt conveys an immediate message of agility and sophistication. Twenty-one-inch wheels and sheer, taut surface relationships reiterate the statement. The Volt's athletic design challenges the notion that an environmentally conscious vehicle can't be beautiful and possess an aesthetic spirit that matches its driving characteristics.

"We leveraged our resources around the globe to develop the design aesthetic for the Volt," said Ed Welburn, vice president, GM Global Design. "It was important that the design capture the face of the Chevrolet as it's recognized around the world." True to the heritage of its Chevrolet bow tie, the Volt's exterior design suggests spirited performance and is wrapped in a stylish package, with classic Chevrolet performance cues that hint at both Camaro and Corvette. On the inside, near-term technologies and innovative materials combine with ingenious use of ambient light for an interior environment that's light, airy and thoughtful.

"First and foremost, this is an advanced technology vehicle that uses little to no fuel at all. But we didn't see any reason why that should compromise its design," said Anne Asensio, executive director, GM Design. Asensio led the design team that created the Volt concept, with designs solicited from GM's studios around the world.

"We wanted a size that connected with everyone, so we designed a small car," said Asensio. "In the end, the interior design team from England inspired the final interior execution, and the exterior is the work of the Michigan advanced design team.

"Our job was to design a vehicle people could easily imagine," said Asensio. "It couldn't be a science project, because that's not what this car is all about. It had to be realistic, executable and carry the essence of the Chevrolet brand." Source: General Motors (GM Europe)

http://www.teslamotors.com Tesla's main web site.

also **http://www.teslamotors.com/blog4/?p=54**

or **http://www.teslamotors.com/blog2/?p=70** for the latest info

Excerpt—Go to this site to read the entire article.

Tesla Roadster Progress—From EP to VP by Malcolm Powell—Vice President, Vehicle Integration published Thursday, March 22nd, 2007

We've just achieved a significant milestone on our road to Tesla Roadster production. Our first Validation Prototype was assembled at the Hethel facility in the U.K. and was recently airlifted to our San Carlos, Calif., workshop to commence system testing.

The Validation Prototypes (VPs) are the second generation of prototypes, succeeding the first generation Engineering Prototypes (EPs). Although many of the EPs are still undergoing testing, VPs incorporate many changes from EP learning. They are much closer in design to the final production vehicles and so enable us to do more refined testing and validation.

July 2010 Newsletter announces the new Roadster 2.5: The new Roadster 2.5 is embodiment of Tesla's commitment to innovation. We unveiled the car last week at the Copenhagen and Newport Beach store openings and on our new website. "These improvements are a direct result of customer feedback and come only a year after release of Roadster 2.0, showing an exceptionally rapid pace of innovation," said Tesla CEO Elon Musk. Roadster 2.5 delivers: more supportive seats, upgraded power control hardware for performance in exceptionally hot climates, and an optional 7" touchscreen display with back-up camera.

Tesla hits the street: Wall Street

Tesla CEO Elon Musk rang the opening bell at NASDAQ in New York City on Tuesday, June 29th in honor of Tesla's initial offering—tthe first for an auto company in more than 50 years.

Look out Government Motors. Here comes a private company.

http://www.marketwire.com/mw Search on Altair or Altairnano

Excerpt—Go to this site to read the entire article.

Apr 23, 2007 14:26 ET

Phoenix Motorcars Exhibits Five-passenger, Zero-emission, All-Electric Mid-Size Sport Utility Truck (SUT) at the Inland Empire Auto Show

Inland Empire Auto Show ONTARIO, CA (MARKET WIRE) April 23, 2007—Ontario-based all-electric vehicle manufacturer, Phoenix Motorcars, will be one of the alternative fuel displays on *Green Street* during the Inland Empire Auto Show scheduled for April 26-29, 2007 at the Ontario Convention Center.

Attendees are encouraged to stop by Phoenix Motorcars' booth to see its five-passenger, zero-emission, all-electric sport utility truck (SUT). During the show, attendees will have the ultimate green experience when they get behind the wheel to test-drive the Phoenix SUT during the Ride and Drive Activity on Saturday and Sunday from 11:00 a.m. to 3:00 p.m.

The SUT can travel at freeway speeds while carrying five passengers and a full payload. It exceeds all specifications for a Type III Zero Emission Vehicle, having a driving range of over 100 miles, can be recharged in less than 10 minutes and has a battery pack with a life of 12 years or more.

A limited number of Phoenix Motorcars all-electric sport utility trucks will be available to consumers in 2007 with an expanded consumer launch scheduled for 2008. Phoenix Motorcars will also introduce an SUV model in late 2007.

Update info September 18, 2009: Details are still very sketchy, but it appears that Phoenix Motorcars filed for Chapter 11 bankruptcy on Monday, April 27. Not surprisingly, the global economic downturn is cited as one of the causes. Apparently, Dubai investor Eqbal Al Yousuf purchased the assets in August of 2009 and will run the company from its California location. They are being very secretive about both development and production plans.

For details see: **http://green.autoblog.com/tag/phoenix+bankruptcy/**

Japan to Set up Public-private Project for Cost-effective Electric Vehicle by 2015; Focus on Lithium-ion Batteries

—*News Release, 7 April 2007* Japan is setting three development targets for advanced battery technologies: improvement (2010), advancement (2015), and innovation (2030). Source: METI Japan's Ministry of Economy, Trade and Industry METI is beginning to organize a research project with business and academia to develop an electric vehicle (EV) by 2015 that costs as much to purchase and to operate as the current generation of minicars, according to a report in the Nikkei.

METI will set up a team comprising researchers from automobile and battery manufacturers as well as universities. The group will be tasked with devising a high-capacity lithium-ion battery that retails for 85% less than existing ones. The new vehicle will likely cost more to purchase than a minicar, but the lower energy cost will make up the difference after about 10 years.

By 2030, METI is targeting the development of a higher-performance battery that can power an electric vehicle for 500 km (310 miles) on a single charge. The ministry will also consider implementing other measures aimed at promoting the use of EVs, including building recharging stations or giving tax breaks to owners of the cars.

The development of electric vehicles is a key element in METI's New National Energy Strategy.

Resources:

Listed below are links to weblogs that reference more information:

Japan to Set up Public-private Project for Cost-effective Electric Vehicle by 2015; Focus on Li-ion Batteries

Recommendations for the Future of Next-Generation Vehicle Batteries (METI)

http://www.meti.go.jp/english and search for Vehicle Batteries

New National Energy Strategy (METI)

http://www.enecho.meti.go.jp/english and search**.**

http://www.greencarcongress.com/batteries/index.html

http://www.greencarcongress.com/japan/index.html

Permalink, Comments (24), Trackback (0)

http://www.greencarcongress.com/2007/04/japan_to_set_up.html

TrackBack URL for this entry: **http://www.typepad.com** is a blogging site. The original information is gone or moved to another place.

Comments:

A 310 mile range by 2030? The Tesla Roadster already has a range of 250 miles using A123 M1 batteries. The Mitsubishi Miex has a range of 150 miles running on current battery technologies.

Will it really take 20 years to double capacity? Given that battery capacity is improving by 9% annually, it will only take 7 years to reach 500 km range. That means 2014. Also, the Miex is set to go on sale for $17,000 in Japan in 2010. So an economical EV by 2015 is rather conservative.

Still good to see Japan taking tech advance seriously. Especially since it includes the cooperation of both public and private entities.

If they can develop the world's most efficient alternative energy tech they will be able to sell it worldwide and revive their stagnant economy.

Posted by: Adam Galas

mail to: **galas010@umn.edu** Apr 7, 2007 10:55 AM

Firefly Energy Awarded Patent on New Foam Lead-Acid Technology; Targeting Hybrids and Plugins, 18 January 2006

Used with permission

Firefly Energy (earlier post) has received a U.S. patent for a new carbon-foam lead-acid battery technology that it believes has the potential to revolutionize the existing global lead-acid battery market as well as serve applications such as hybrid electric vehicles and plugin hybrids. Firefly contends it can deliver lead-acid battery performance comparable to NiMH, but at about one-fifth the cost, and with much reduced weight compared with traditional lead-acid batteries. As a result, the company believes it can play an important role in accelerating the adoption of hybrid and plugin hybrid vehicles through cost reduction and availability.

The Firefly battery replaces the conventional lead plates in a lead-acid battery with a lightweight carbon or graphite foam to which the chemically active material in the form of a paste or slurry has been applied. The use of the foam structure increases the interface between the electrodes and the active chemistry; the carbon material resists corrosion and sulfation build-up, reducing weight and delivering a formidable jump in specific power, energy and cycle life. The technology is not limited to use in lead-acid batteries.

Firefly is a spin-off from Caterpillar, which had assigned the problem of pursuing increased performance for lead-acid batteries used by Caterpillar's product groups to Kurt Kelley, who is now Firefly Energy's Chief Scientist. Since Kurt, an accomplished material scientist, had never designed a battery before, his problem-solving approach was unconstrained by the conventional battery wisdom held by lead acid battery technologists.

Edward Williams, CEO and Firefly Energy cofounder Kelley came up with the idea of using a foam carbon composite to address the corrosion and sulfation issues, and to remove the bottlenecks to achieving the theoretical power of the lead acid chemistry.

A major restriction to lead-acid battery efficiency is the lack of interface area between the active chemistry and the electrodes. Although the chemistry is theoretically capable of delivering approximately 170 Watt Hours per Kilogram (Whr/kg), lead-acid batteries only average around 30 Whr/kg.

Up to now, achieving a higher surface area within a given lead-acid battery box required the addition of more and thinner lead electrodes. However, lead electrodes corrode, so increasing surface area by putting thinner lead electrodes in the battery increases corrosion and decreases battery life.

Removing the corrosive heavy lead grids and replacing them with a graphite foam addresses both issues (increased surface area and decreased corrosion). Furthermore, the design of the Firefly battery removes one-half to two-thirds of the lead out of the battery.

Firefly Energy is now beginning to promote the use of its foam lead-acid batteries in plugin hybrid applications.

Senior Vice President Mil Ovan most recently made that pitch at the Energy Independence 2020 Summit organized by United States Senator Dick Durbin (D-IL) in Chicago.

New information on the Firefly Oasis battery:

http://thefraserdomain.typepad.com/energy/2007/10/firefly-truck-b.html

Firefly Truck Battery to be Available for Evaluation in First Quarter 2008

Per press release, October 30, 2007:

Firefly Energy Inc. the Peoria, Illinois based leader in developing next generation carbon and graphite foam batteries, announced today that the first preproduction versions of its BCI Group 31 truck battery—to be marketed under the new name *Oasis*—will be available for review and testing during the first quarter of 2008.

The company said its up to 50 percent longer runtimes than competitors when Oasis battery will primarily be utilized when the truck's diesel engine is turned off, and provide powering accessories which collectively make up a truck's *hotel loads*. This newly branded battery will be unveiled by Firefly Energy at its first ever trade show appearance at the SAE Commercial Vehicle Engineering Congress and Exhibition (ComVec), Oct. 30-Nov. 1, 2007, in Rosemont, IL.

"Anti diesel engine idling regulations will soon become pervasive across the nation, and better battery performance will be crucial in contributing to trucker safety, comfort, and productivity," said Ed Williams, chief executive officer of Firefly Energy. "Our Oasis battery will help ensure truck drivers maintain a comfortable haven for their rest periods"

For more information on Firefly go to:

http://www.fireflyenergy.com/

http://www.startribune.com/world/   The original article is no longer available at its previous address, but might be found by searching.

Chinese battery firm rolls out mass-produced hybrid car

excerpt—By DON LEE, Los Angeles Times: December 13, 2008

SHENZHEN, CHINA - With the Detroit Big Three automakers teetering and China's once go-go car market in reverse, this might seem a bad time for a relative unknown to be launching a new vehicle. Then again, BYD Co. is not rolling out any ordinary car.

On Monday, the upstart company best known for making cell-phone batteries will begin selling its F3DM -- China's first mass-produced hybrid electric vehicle. The car is expected to retail for around $20,000 in China and make its way to the United States in 2011.

"For a long time, China's auto technology was undeveloped," Wang Chuanfu, the 42-year-old founder and president of BYD, said in an interview Friday at its headquarters here. "But our [electric car] technology marks the first time we're standing as a leader on the world stage."

Whether that assessment is overblown -- Toyota, Honda and General Motors may certainly think so -- people familiar with Wang say they wouldn't underestimate him. Since starting the company in 1995, Wang has built BYD, short for Build Your Dream, into the world's leading producer of rechargeable batteries for mobile phones and laptops, among other products. Now joining with Daimler to produce electric cars for the Chinese market could be a big positive move. See this site for latest news.

http://www.allcarselectric.com/blog/1043629_byd-and-daimler-join-for-work-on-evs-for-chinese-market

---

http://findarticles.com/p/articles/mi_m0EIN/is_/ai_59321714

Power Technology Announces Advances With Douglas Battery Relationship

excerpt—LAS VEGAS--(BUSINESS WIRE)--Feb. 10, 2000

POWER TECHNOLOGY, Inc. (OTC BB: PWTC), a Las Vegas technology development company, announced today that developments with Douglas Battery (www.douglasbattery.com) are progressing.

Power Tech and Douglas Battery are in active discussions on technical assistance, disposition of rights of technical information, processes and products developed.

Power Tech's ultimate desire is to optimize its environmentally friendly nickel-ferrous battery to compete head-to-head with lead acid, nickel metal hydride, lithium-ion, and zinc-air, on a power density, weight, environmentally friendly, life expectancy and cost-per-watt-hour basis.

http://autos.yahoo.com/green_center

Excerpt—the entire article is no longer posted on this website, but there is a great deal of information available on this site.

Back from the Dead? The Future of Electric Vehicles.

Jon Alain Guzik, Editor-at-Large, Yahoo! Autos:

The Honda FCX is a mix of carbon fiber construction, bleeding-edge technology and a smaller fuel-cell stack that turns hydrogen into electricity, which in turn drives the FCX's 95kW AC synchronous motor. The FCX on the track is silent—with just the sound of the motor spooling up—and quick, with all the torque delivered at low rpms. A bonus: the only emission coming out of the tail pipe is good old H_2O, which looks cleaner than the water that comes out of the faucet in my own home.

Author's note: That may be true, but all that has been done is the transfer of emission of carbon dioxide from the exhaust pipe of the vehicle to the exhaust stack of a power plant. Since it takes more energy to make hydrogen than can possibly be extracted by any fuel-cell, the result will probably be even more carbon dioxide in the atmosphere.

While this may all seem perfect, hydrogen being a big buzzword these days, a few obstacles stand in its way, namely, the infrastructure to produce and deliver hydrogen that powers these types of vehicles. A few other hurdles besides infrastructure: the size of the fuel-cell and the power train package, the short mileage range and cold weather durability. But look at Honda's first FCX from 2002 and the new model side-by-side and you can see the giant steps taken in the last six years alone.

"Honda's position is that there is not one silver bullet to the solution. We see a hydrogen fuel-cell as part of the solution—zero emissions and in theory, in unlimited sustainability," says Sage Marie, a Honda spokesman. When asked what he predicts is coming in the next decade, he replies, "Hydrogen is our vision and it will become a larger part of the model mix than today. Ten years from now, I'd expect more diesels in the lineup as well as more hybrids. You'll see an expansion of today's alternative fuels like diesel and natural gas. In our vision, hydrogen will be a part of that mix."

http://autos.yahoo.com/green_center

Excerpt—The entire article is no longer posted on this website.

The EVs of Today and Tomorrow.

A large portion of the electric vehicles on the road today are what's called Neighborhood EVs, meant to be driven at speeds under 25 miles an hour. But future highway-speed vehicles are coming soon due to high consumer demand. Why? A large portion of the buying public these days are looking for a change from their gas-powered cars and only drive from 25-50 miles a day to and from work, a perfect scenario for an EV vehicle.

Also, they look pretty darn cool. It's the coolness factor that makes people take a closer look at what electric vehicle technology has to offer because these vehicles aren't very fast, are expensive and, without a long driving range or an easy way of refilling the batteries, somewhat impractical.

Author's note: Not all EVs fit the description in the first paragraph. New battery technology is showing promise of even longer ranges to come as well as realistic prices. Toyota's RAV4 EV from a few years back proved that electric vehicles were not only practical but economical and very desirable as well. Why Toyota stopped making them and tried to destroy all those that had been made is a subject for much conjecture just like GM's Chevy S10 EV. The Phoenix EV is already on the market and the waiting lines are long for this pickup with a top speed of more than a hundred and working range of about 200 miles.

Some of the new vehicles described include, **Phoenix Motors SUED, GM Volt, Myers Motors NMG, and AC Propulsion eBox.**

This list does not include several other EVs and none of the PHEVs now appearing or shortly to appear on the market. There are indeed a promising number of alternatively powered vehicles—vehicles that use no petroleum as fuel, being introduced by American entrepreneurs at the present. The innovative American and American company have not yet vanished from the scene.

Neighborhood electric vehicles—A niche market

Two web sites:

http://www.zapworld.com/about-us

http://en.wikipedia.org/wiki/Neighborhood_electric_vehicle

While many do not truly qualify as cars, neighborhood electric vehicles (NEVs) are becoming more common and numerous. Mostly based on golf carts or similar conveyances, these specialty vehicles have limited range, limited speed and so limited use. Available in open and closed types their speed is no more then 30 mph and their ranges from 30 to fifty miles. There are some exceptions to both as speeds and ranges can be increased with new types of batteries. Several automotive companies developed these types of vehicles a few years ago, but never brought them to market. Ford was one of these with their EV1. Time will tell whether these niche-market vehicles will gain popularity or not. They certainly are looking more attractive for people who drive close to home, particularly as the price of gasoline roars past four dollars a gallon with no ceiling in sight. All that is needed under these conditions is for a manufacturer to design and build a practical NEV and do a good job of marketing. It would require a substantial change in our thinking about how we get around for work and shopping, but that is certainly not impossible as fuel costs soar. Economic pressures could produce what the tree huggers could never accomplish.

http://www.ev1.org/msg/19.htm

NiMH: Nickel-metal Hydride Batteries—why are they not being used?
Some questions from one who would like us all to know.

Used with permission

What's going on here? Is GM truthful about NiMH and Chevron owning the patent rights? Why did Californians have to mount a *don't crush* campaign to stop Toyota from recalling and crushing nearly a hundred electric RAV4 EVs in Santa Monica?

Will calling for GM to produce the Volt with NiMH batteries be impossible, unless GM goes to Chevron-cobasys, to which it sold the patent rights, and begs to be able to use the batteries?

Why does this force GM to refuse to even SEE the existing hundreds of Toyota RAV4-EVs. Maybe it forces them to admit that they can't get the batteries from Chevron!

Now, is Toyota being truthful? One of their honchos said that "the batteries are not ready" on April 3, 2007, even while in sight of many Toyota RAV4 EVs using the same NiMH batteries.

A curious fact, that the auto companies refused to allow even ONE plugin EV, still proclaiming that the GM *fleet* of 100 fuel-cell cars, to be released this year, is the way of the future. These promised fuel-cell cars have fewer fueling stations even than ethanol or biodiesel cars; and they are only limited to 50,000 miles of life.

Author's note: The author is convinced that the hydrogen fuel-cell system is being foisted off on the unsuspecting American public by people who know it will never be economically feasible, because they know it will never be economically feasible. This is being done for the purpose of keeping other technologies that are economically viable from getting any support, financing or publicity. Considering these people generously, they have the same kinds of views of the future as the people who tried to stop Eli Whitney's cotton gin and fought against automation in our factories.

Several far more sinister motivations could be at work here. These include serious efforts to stifle competition by one of the largest and most powerful groups of multinational corporations in league with many of the world's most despotic regimes. These people love the ruling power of petroleum and will do virtually anything to thwart the efforts of any serious competitor. This is particularly true of any competitor with a real chance to capture the energy market with a usable liquid fuel and fueling system that costs less than gasoline or diesel.

Who's kidding whom? The *equinox* GM cars cost about $1 million each, and have less range than the Toyota RAV4-EV, cost 20 times as much as the Toyota RAV4-EV, last less than a fourth the life of a Toyota RAV4-EV, and require $1 million hydrogen fuel stations, while the Toyota RAV4-EV can charge off a regular PLUG! Ask GM why? Ask Toyota why? There was certainly nothing wrong with these experimental vehicles. They performed better than expected.

March 16th, 2007 at 8:43 am

When GM showed the inner workings of their hybrid battery research lab, why did no reporter even ask GM why they do not use the cheaper, longer lasting, well-proven, still-running Nickel Metal Hydride (NiMH) batteries used in the 1999 EV1, Ranger-EV, Honda EV-plus and still in use in the 2002 Toyota RAV4-EV?

In fact, NiMH is the only production EV battery used in all plugin cars that attained an all-electric range of more than 100 miles for more than 100,000 miles.

According to an assessment by the California Air Resources Board, NiMH costs from $225 to $350 per kWh, no more than $13,000 for a large pack for a full EV, including retail profit and all components. The Volt would only need a NiMH pack a third that size; moreover, part of the battery cost would be covered by not having to install a transmission, clutch, etc.

Lithium-ion cells cost $1000 per kWh and up for batteries represented to not have the thermal runaway problem, and even the riskier laptop batteries, with all the economies of mass production of hundreds of millions of batteries, cost more than $400/kWh.

NiMH does not require expensive research toward an uncertain goal that may never be found; but a Lithium-ion wild goose chase does allow GM to divert attention from the already available plugin EV cars.

Why not ask GM's Bob Lutz: "Why aren't you using the cheaper, longer lasting, most-tested, standard Electric car battery pack, the only battery proven to last longer than the life of a car?"

Doug Korthof 562-430-2495

Author's note: *The rapidly advancing technologies described elsewhere in this appendix may hold the answers to these questions. Sadly, we may never know the real truth about the end of the EV programs. Many suspicions about things deliberately kept from the public are described in the following website information:*

http://ev1.org/index.htm#Yakyak

The original site from which this was excerpted is gone. Check
http://www.gizmag.com/go/7291/ or look up Chevrolet Sequel on Google.

Excerpt—Go to this site to read the entire article.

Chevy Sequel: GM Press Release

TARRYTOWN (May 15, 2007) General Motors Corp. made history today as its Chevy Sequel—the world's most technologically advanced automobile—became the first electrically-driven fuel-cell vehicle to achieve 300 miles on one tank of hydrogen, in and out of traffic on public roads, while producing zero emissions.

"With this drive, General Motors has reached another important milestone toward the commercialization of our fuel-cell vehicles, by achieving the range expected by today's consumers," said Larry Burns, GM vice president, research and development and strategic planning. "And we did it while producing zero emissions, as a hydrogen fuel-cell vehicle only emits water. In addition, the hydrogen produced at Niagara Falls, used to fuel Sequel, was derived from hydro power—a clean, renewable resource. This means that the entire process—from the creation of the hydrogen to the use of the fuel in the vehicle—was virtually carbon dioxide free."

Sequel was introduced in 2005 at the North American International Auto Show in Detroit and the first driveable version appeared last fall. It is the first vehicle in the world to successfully integrate a hydrogen fuel-cell propulsion system with a broad menu of advanced technologies, such as steer—and brake-by-wire controls, wheel hub motors, lithium-ion batteries and a lightweight aluminum structure. It uses clean, renewable hydrogen as a fuel and emits only water vapor. Now it is the first fuel-cell vehicle to achieve real-world range.

Author's note: *It may be true that it emits only water vapor, but all that has been done is the transfer of emission of carbon dioxide from the exhaust pipe of the vehicle to the exhaust stack of a power plant. Since it takes more energy to make hydrogen than can possibly be extracted by any fuel-cell, the result will probably be even more carbon dioxide in the atmosphere.*

http://www.thecarconnection.com/blog/?p=487

Used with permission from TheCarConnection.com

Is GM Putting Too Much on Batteries?

General Motors is charged up about electric power, but will the lack of the right batteries short-circuit high-profile projects, such as the Chevrolet Volt, the plugin hybrid that garnered so much attention at the Detroit Auto Show earlier this month?

Okay, it's easy to get caught up in all the punning, but the subject is deadly serious and could affect the long-term transition away from petroleum, never mind the competitive nature of the auto industry.

Let's face it: the giant automaker has a credibility issue. It's no longer seen as the industry leader in many communities, and that includes many lawmakers, rule-makers and other opinion leaders. Many now look to GM's arch-rival, Toyota, to set the pace when it comes to environmentally-friendly automobiles. So the Detroit maker had a lot riding on the roll-out of the Volt, which it claims could let most U.S. commuters go to work solely on battery power. For longer drives, the Volt would still be able to fire up its internal combustion engine. There's only one problem, cautioned GM's Bob Lutz: the batteries necessary to make the Volt work aren't a commercial reality yet. That same caveat out of Toyota would be readily accepted. From GM, though, many skeptics are left wondering whether it's just another excuse.

So I was curious to see what the company would have to say at a Monday morning battery briefing at the GM Technical Center, in Warren, Mich. An assortment of General Motors officials were joined by counterparts from the supplier community, including Mary Ann Wright, former head of the hybrid program at Ford and now the CEO of Johnson Controls-Saft. JCS is a joint venture of the U.S. supplier and the French battery maker and it's pushing hard to commercialize lithium-ion technology, which seems to be the breakthrough everyone needs. "It's not a revolution, but an evolutionary challenge facing GM and its competitors." insists Joe LoGrasso, the battery boss at the Tech Center. "But there are some real hurdles. Just consider all those exploding lithium-ion laptop computer batteries that have been making the headlines lately. Would you like one of them powering your $35,000 SUV?"

But is GM setting the benchmark too high? The automaker says it will not buy in on lithium technology unless and until battery makers, like JCS, can ensure safe and reliable technology that can also last the life of the car, meaning at least 100,000 miles without replacement. Can you imagine your clutch or perhaps even a transmission going that long without repair or replacement? Is GM setting a target that it knows can't be met? And will competitors like Toyota accept a lower standard to beat the U.S. maker to market? That would be the worst of all worlds and yet another devastating blow to the domestic giant's credibility. General Motors desperately needs to take a solid leadership role in the green car movement and simply fielding concept cars, even those as promising as the Volt, won't cut it.

Wrapping up the session, GM officials suggested they might have the necessary lithium-ion technology ready for prime time by 2010. They may not have that much time.

Author's note: Perhaps GM officials should take a look at other battery technologies. These batteries could be better and available earlier than those using lithium ion technology. Maybe they should just read the article that follows. Of course, now that GM is Government Motors and most of their top people (including Lutz) have fled from government bureaucracy and UAW control, I doubt we will see much innovation or profit.

http://www.nanotechwire.com/news.asp?nid=3753&ntid=&pg=57

Altair nanotechnologies achieves breakthrough in battery materials. Altair's Developments pave the way for a new generation of rechargeable batteries.

NEWS RELEASE

9/15/2006 6:13:50 PM

Altair nanotechnologies details power features of its nano titanate battery.

Altair Nanotechnologies Inc., a leading provider of advanced nanomaterials and alternative energy solutions, detailed why its NanoSafe rechargeable, nano titanate battery technology provides fundamental improvements, including high power versus other rechargeable batteries.

In anticipation of Altairnano's delivery of its first NanoSafe battery pack for use in an electric vehicle in September; this is the final of four planned news releases identifying features of Altairnano NanoSafe batteries that may prove advantageous in the power rechargeable battery market. In the three previous releases, Altairnano detailed why its nano titanate battery technology delivers high battery safety, rapid recharge and long battery life. The combination of these features has the potential to make Altairnano's NanoSafe batteries ideal for power applications such as electric vehicles and hybrid electric vehicles.

How Does a Rechargeable Battery Work?

A battery consists of a positive electrode, a negative electrode, a porous separator that keeps the electrodes from touching, and an ionic electrolyte, which is the conducting medium for ions (charged particles) between the positive and the negative electrodes. When the battery is being charged, ions transfer from the positive to the negative electrodes via the electrolyte. On discharge these ions return to the positive electrode releasing energy in the process.

Existing Lithium-ion Batteries

Rechargeable lithium-ion batteries currently use graphite for the negative electrode and typically lithium cobalt oxide for the positive electrode. The electrolyte is a lithium salt dissolved in an organic solvent which is flammable.

An important attribute of large format batteries is their ability to deliver power quickly. During charge, lithium-ions deposit inside the graphite particles. However, the rate at which lithium-ions can be removed during discharge—the useful power-producing cycle of a battery—is limited by the electrochemical properties of the graphite and the size of the graphite particles. The electrochemical properties relate to the existence of a high resistance crust (call the Solid Electrolyte Interface or SEI) that impedes the removal of lithium—the first step in power production. Also, graphite's large particle size means that lithium atoms inside the particle must travel a long distance to escape. This further increases the impedance and reduces power.

So power is restricted by the ion removal capability in lithium-ion batteries, resulting in power levels of the order of 1000 watts per kilogram (W/Kg). Also, power can be affected by external factors such as temperature. At low temperatures, the lithium-ion removal rate is significantly less than at room temperature resulting in power delivery at these temperatures that is substantially reduced.

Given that power delivery is governed by fundamental properties of the materials the only option is to change the materials and chemistry of the battery.

The Altairnano NanoSafe Battery

Altairnano solved this problem by using an innovative approach to rechargeable battery chemistry by replacing graphite with a patented nano titanate material as the negative electrode in its NanoSafe batteries. The outcome is that Altairnano's NanoSafe batteries deliver power per unit weight and unit volume several times that of conventional lithium-ion batteries. Altairnano laboratory measurements indicate power density as high as 4000 W/Kg and more than 5000W/liter. By using nano-titanate materials as the negative electrode material, the formation of an SEI is eliminated.

In addition, the nano-titanate particles are up to 100 times smaller than a typical graphite particle thereby reducing the distance a lithium atom must travel to be released from the particle. These properties also mean that even at cold temperatures, a nano-titanate battery will produce high power.

The same technology also dramatically increases battery charge and discharge rates; rapid charge is important for next generation electric vehicles so they could be charged in a few minutes rather than hours as with current lithium-ion technology. As has been indicated in previous releases the NanoSafe cell has demonstrated that surges of power can be delivered without risking thermal runaway or performance damage to the battery.

Altairnano will be demonstrating its NanoSafe battery technology at the California Air Resources Board Zero Emission Vehicles meeting in Sacramento, September 25th through 27th, 2006.

More from Altair: The original site is gone, instead go to:

http://www.marketwire.com/mw and search on Altairnano.

RENO, NV (MARKET WIRE) February 10, 2005

Altair Nanotechnologies, Inc. (NASDAQ: ALTI) announced today that it has achieved a breakthrough in Lithium-ion battery electrode materials, which will enable a new generation of rechargeable battery to be introduced into the marketplace, as well as create new markets for rechargeable batteries. These new materials allow rechargeable batteries to be manufactured that have three times the power of existing Lithium-ion batteries at the same price and with recharge times measured in a few minutes rather than hours.

The technical achievements are being praised by the battery community as truly remarkable and will likely enable a new generation of rechargeable battery to be produced. Altair has confidentiality agreements in place with some of the world's leading battery development companies to evaluate and commercialize these battery electrode materials.

Altair's research and development efforts were allowed two new patents (announced on January 7th and 14th, 2005) and a National Science Foundation grant was successfully completed in January, 2005, by Altair. New markets for fast charging batteries will include the handheld power tools market increasing the productivity of, for example, construction workers while lowering their overhead costs. Other markets include hybrid electric vehicles, portable electronics and medical surgery tools—solving the problem of electrical wires all over the operating room floor.

Endorsements for Altair's work in this area have come from many quarters. Two eminent experts in battery technology, Dr. K. M. Abraham and Dr. Vassilis G. Keramidas, have expressed strong support for Altair's work "The nanomaterials Altair is developing are the next generation of electrode materials for lithium-ion batteries and Altair's research and product development is laying the ground work for a new generation of ultra high power lithium-ion batteries," commented Dr. K. M. Abraham. "A key requirement to the above applications is the ability to recharge the battery quickly, for example in a few minutes. Current Li Ion batteries are incapable of such quick charge times because of the chemistry of the anode materials. Altair has found a solution to this with their nano-sized lithium titanium oxide."

"Altair's nanomaterials, which have a virtually zero strain crystal lattice, eliminate the main cause for battery electrode material fatigue, which limits rechargeable battery life, increasing the number of recharge and discharge cycles from a few hundred to many thousand cycles," said Dr. Vassilis G. Keramidas. "I find Altair's development strategy and proposed research direction sound and a necessary step in establishing the Li-ion electrochemistry as a viable contender for large battery applications."

"Our research in battery electrode materials is a further indication of how Altair's scientists are able to apply their nanomaterials science knowledge to solving real world needs," commented Altair CEO Dr. Alan J. Gotcher. "Many of the technology and product development initiatives that we have been working on for the last few years are now coming to the commercialization stage. Each step is another validation of our business strategy and product technology platform. Altair's nanomaterial based, micro porous electrode technology has performance and stability advantages that appear to be unmatched when compared with the best commercialized technology in the market today."

Dr. K. M. Abraham

Dr. Abraham, the principal of E-KEM Sciences, has more than 28 years of experience in lithium-ion battery research and development. He has made pioneering contributions to the development of the rechargeable battery industry. Dr. Abraham has authored and coauthored more than 150 publications and 16 U.S. patents in battery technology. He received his PhD from Tufts University and conducted postdoctoral research at Vanderbilt University and MIT. In 2000 he was elected a Fellow of the Electrochemical Society. Dr Abraham is on Altair's Scientific Advisory Board.

Dr. Vassilis G. Keramidas

Dr. Keramidas, Managing Director of Keramidas International, provides management consulting to major corporations on the formulation and execution of their research and development. Prior to founding Keramidas International, he was Vice President of Applied Research at Telcordia Technologies (formerly Bellcore). His division at Telcordia made seminal contributions in the fields of photonic and electronic materials. The Bellcore research on energy storage ushered in the polymer battery Li-ion technology with the invention of the Bellcore Plastic Li-ion (PLiON) technology. Dr. Keramidas sits on the Board of two battery companies.

Altair Nanotechnologies, Inc. Through product innovation, is a leading supplier of advanced ceramic nanomaterial technology worldwide. Altair Nanotechnologies has assembled a unique team of material scientists who, coupled with collaborative ventures with industry partners and leading academic centers, has pioneered an impressive array of intellectual property and product achievements.

Altair Nanotechnologies has developed robust proprietary technology platforms for manufacturing a variety of crystalline and noncrystalline nanomaterials of unique structure, performance, quality and cost. The company has a scalable manufacturing capability to meet emerging nanomaterials demands, with capacity today to produce hundreds of tons of nanomaterials.

The company is organized into two divisions: Life Sciences and Performance Materials. The Life Sciences Division is pursuing market applications in pharmaceuticals, drug delivery, dental materials, cosmetics and other medical markets. The Performance Materials Division is pursuing market applications in Advanced Materials for paints and coatings;

titanium metal manufacturing, catalysts and water treatment; and alternative energy. For additional information on Altair and its nano-materials, visit www.altairnano.com.

Altairnano™, Altair Nanotechnologies, Inc.®, Altair Nanomaterials™, TiNano®, RenaZorb™, NanoCheck™, TiNano Spheres™ and the Hydrochloride Pigment Process™ are trademarks or registered trademarks of Altair Nanotechnologies, Inc.

Names to contact for additional information:

Marty Tullio or Mark Tullio

McCloud Communications, LLC—949.553.9748

marty@mccloudcommunications.com

mark@mccloudcommunications.com

---

http://www.a123systems.com/html/news/articles/012507_pr.html

Excerpt—Go to this site to read the entire article.

A123systems receives $40 million investment to expand product portfolio and scale manufacturing of hybrid and plugin hybrid batteries.

Procter & Gamble's Duracell and A123Systems Collaborate on Future Products

Watertown, Mass. January 25, 2007—

A123Systems, one of the world's leading suppliers of high-power lithium ion batteries, today announced it has completed a $40 million round of funding, bringing the total capital invested in the company to $102 million. A123Systems will use these funds to scale its technology development and manufacturing capacity for plugin hybrid electric vehicle (PHEV) batteries, as well as to support the fast-growing demand in the power tool, hybrid electric vehicle (HEV) and consumer applications markets.

"As our doped nanophosphate technology continues to deliver superior power, safety and life, we're experiencing tremendous growth and interest from companies across a vast array of industries looking to innovate new and existing products," said David Vieau, CEO and President of A123Systems.

For additional information please visit: **www.a123systems.com**.

Contact: Keith Watson, Fama PR, 617-758-4154 **keith@famapr.com**

Author's commentary: Why nanotechnology could be the biggest payoff since the advent of the steam engine.

The new industrial revolution, like the one that occurred in the early 1900s, is a vibrant, and significant developing phenomena with both economic and political implications of change. This new technology is invading virtually all parts of every industry in the world. Like the bioindustries, nanotechnology involves things too tiny to be seen with the naked eye. While everyone stands to benefit from this new wave of applied technology, no one will ever actually see it.

The new industrial revolution I'm talking about is nanotechnology, the precision creation and manipulation of matter on the atomic scale. This sounds futuristic, I know, but nearly a thousand companies worldwide are already involved in nanotechnology. In the last year alone, corporations and governments worldwide have pumped several billions into research and development in this exciting new sector. What is more important, companies have already applied this technology to a variety of consumer products, including automobile parts, semiconductors, clothing, sports equipment and toys, to name just a few.

You've probably already read something about the developments occurring in nanotechnology. Yet most people have no exposure to this fast moving and rapidly growing part of the world's economy.

The global market for nanotechnologies could reach $1 trillion dollars or more within the next twenty years. Nanotechnology could make crude oil obsolete in a few years with new battery technology for all-electric cars.

http://ec.europa.eu/research/press/2002/pr1206en.html

Excerpt—Go to this site to read the entire article.

Brussels, 12 June 2002

Nanotechnology, Information Technology, Industrial Processes:

The nanotechnology revolution has started. At the cutting edge of science and innovation, nanotechnology offers unprecedented challenges and opportunities for researchers, businesses, and investors in Europe. Already fueling innovative applications in industries as diverse as IT, automotive, cosmetics, chemicals, and packaging, nanotechnology also holds considerable promise to generate radical new applications—and foster the development of whole new sectors of activity. Amongst the most promising are energy storage and distribution; detection, measurement and testing; processors, bio-analysis and drug delivery, robotics and prosthetics.

In this context, the EU organizes an information event on Nanotechnology: a New Industrial Revolution at CEA-Minatec, in Grenoble, France, on 14 June 2002. This information event will immediately follow on a major EU-US conference on nanotechnology and nanomanufacturing, the third in a series of conferences held in conjunction with the U.S. National Science Foundation, also held at CEA-Minatec. Specifically aimed at the press, it will gather some of the best specialists from research, industry and finance on both sides of the Atlantic. It will address the key scientific, technological, and economic challenges of nanotechnology, and highlight the opportunities for Europe's researchers and entrepreneurs.

NOTE: Further information on nanotechnology is available from Herv Pero, Head of Unit, Products, Processes and Organization, Directorate-General Research,

Tel. + 32 2 298 1232 E-mail **herve.pero@ec.europa.eu**

http://www.b-eye-network.com/view/191

The Nanotechnology Revolution by Dan E. Linstedt

Used with permission

Published: June 30, 2004

The nanotechnology revolution has arrived! Business intelligence and data warehousing need to prepare for the *atomic* age.

There is a lot of buzz in the manufacturing world these days all about Nanotechnology. It has the ability to repair, repel, construct, destruct and detect different chemical elements down to the core of atoms. The government has more than a few initiatives in this sector, directed by a tax-funded group called DARPA (Defense Advanced Research Projects Agency, www.darpa.mil). However, in the commercial sector of data warehousing and business intelligence not a lot of attention has been paid to Nanotechnology. This is unfortunate, as Nanotechnology may be the next revolution affecting all parts of life down to the atomic-level.

This article attempts to peer over the very edge of this technology. It examines theories of what might be headed our way. This is particularly true if this technology is applied to information stores we commonly know as data warehousing. This is a hypothetical look at some of the real Nanotechnologies already available, and how data warehousing and business intelligence might change.

What is nanotechnology?

An oversimplified definition: Nanotechnology is the ability to represent information encoded at a molecular or atomic-level. For example, one particular use of these encoded molecules (when appropriately arranged) is to function as a superconductor molecular *wire* or tubule. Nanotechnology is the application of arranging molecules to perform specific tasks on an atomic-level. When multiple layers of these molecular items are combined, it becomes possible to conceive macro-level applications of these creations—such as stain-free pants or radio frequency identification (RFID) tags.

No one will be immune to the changes created by Nanotechnology. Simply run a search on Amazon, Yahoo, or Google and you will see the vast number of companies and the books being written to address this new technology.

Where is Nanotechnology used today? An example of the macro-level applications is the pants that resist staining. Another example may be self-sealing tires; the tires that when punctured respond and patch the puncture automatically. There are many different emerging applications of Nanotechnology. RFID (Radio Frequency Identification) is an example of a widespread application.

There are a number of different technologies that have already been demonstrated for communicating in both directions between the wet, analog world of neurons and the digital world of electronics. One such technology, called a neuron transistor, provides this two-way communication.

For more on nanotechnology, but in the same vein see:

http://www.kurzweilai.net/news/frame.html?main=/news/news_single.html/?id%3D7 948

If this site is no longer active goto: **http://www.kurzweilai.net** and search on *nanotechnology*.

Ray Kurzweil—*The Human Machine Merger:* Are We Headed for the Matrix?

http://www.buckypaper.com

Buckypaper, New Nanotechnology Development with Promise.

Excerpt—Some late breaking information on nanotechnology:

TALLAHASSEE, Florida (AP)—It's called *buckypaper* and looks a lot like ordinary carbon paper, but do not be fooled by the cute name or flimsy appearance. It could revolutionize the way everything from airplanes to TVs are made.

Buckypaper is 10 times lighter but potentially 500 times stronger than steel when sheets of it are stacked and pressed together to form a composite. Unlike conventional composite materials, though, it conducts electricity like copper or silicon and disperses heat like steel or brass.

"All those things are what a lot of people in nanotechnology have been working toward as sort of Holy Grails," said Wade Adams, a scientist at Rice University. That idea—that there is tremendous future promise for buckypaper and other derivatives of the ultra-tiny cylinders known as carbon nanotubes—has been floated for years now. However, researchers at Florida State University say they have made important progress that may soon turn hype into reality.

Bucky paper is made from tube-shaped carbon molecules 50,000 times thinner than a human hair. Due to its unique properties, it is envisioned as a wondrous new material for light, energy-efficient aircraft and automobiles, more powerful computers, improved TV screens and many other products.

So far, buckypaper can be made at only a fraction of its potential strength, in small quantities and at a high price. The Florida State researchers are developing manufacturing techniques that soon may make it competitive with the best composite materials now available.

"If this thing goes into production, this well could be a major game-changing or revolutionary technology to the aerospace business," said Les Kramer, chief technologist for Lockheed Martin Missiles and Fire Control, which is helping fund the Florida State research.

Typically the original New York Times article excerpted here is gone, but the reader can get the main message.

Ex-Environmental Leaders Tout Nuclear Energy By MATTHEW L. WALD, *New York Times*: April 25, 2006

"The nuclear industry has hired Christie Whitman, the former administrator of the Environmental Protection Agency, and Patrick Moore, a cofounder of Greenpeace, the environmental organization, to lead a public relations campaign for new reactors."

Should you want to read the entire article, try to search the *New York Times* site or in the following: **http://www.ens-newswire.com**

U.S. Nuclear Industry Fires Up Public Relations Campaign

By J.R. Pegg

WASHINGTON, DC, April 24, 2006 (ENS)—The nuclear industry launched a new campaign on Monday to generate support for increased nuclear power, spearheaded by Greenpeace cofounder Patrick Moore and former U.S. Environmental Protection Agency Administrator Christine Todd Whitman.

Nuclear power advocates are hoping that Moore and Whitman can sell the American public on the benefits of nuclear power and help spark the resurgence of an industry that has not constructed a new plant in some 30 years.

"Scientific evidence shows that nuclear power is an environmentally sound and safe energy choice," said Moore, who along with Whitman will co-chair the Clean and Safe Energy Coalition (CASEnergy).

Christine Todd Whitman is a former governor of New Jersey, and the first head of the U.S. Environmental Protection Agency under President George W. Bush. The coalition, which is funded by the Nuclear Energy Institute (NEI), includes more than 50 charter member organizations, including the U.S. Chamber of Commerce, the International Brotherhood of Teamsters and the National Association of Manufacturers.

NEI represents the owners and operators of the nation's 103 commercial nuclear reactors—these facilities currently produce 20 percent of U.S. electrical power.

Jim Riccio, a nuclear power analyst with Greenpeace USA, said Moore has been "living off his reputation with Greenpeace for some time now and lacks credibility. To call nuclear power clean and safe is the height of hypocrisy, especially as we are ready to commemorate Chernobyl," Riccio told ENS.

Wednesday is the 20th anniversary of the Chernobyl disaster in Ukraine—the world's worst nuclear power accident. Although U.S. plants are much safer than the doomed

Chernobyl facility, critics remain unconvinced that the nation's regulatory agency, the Nuclear Regulatory Commission, or the nuclear industry, in fact focus on safety. A report released Monday by Greenpeace finds that the industry has had some 200 *near misses* to nuclear meltdowns since 1986.

"The study shows that nuclear power plants are a *clear and present danger*," Riccio said, "and packaging nuclear power as a solution to global warming is *dead wrong*. The primary driver of increasing U.S. carbon dioxide emissions is the transportation sector," he said, "and nuclear power will do nothing to address the nation's thirst for oil."

Author's note: *Mr. Riccio would do well to stick to facts and avoid his fearmongering alarm tactics. The fact is the newest light water reactors are far less dangerous over all than coal-fired power plants. Another salient fact is that the nuclear power industry has a far better health and safety record than all others as reported earlier in this book in the section on nuclear power. As of 2004, and since 1972, fatalities directly related to coal-fired power plants per terawatt of power produced numbered 342. Hydroelectric fatalities were 883, and natural gas 85. Only eight fatalities were recorded for the same terawatt of nuclear power. Add to these statistics the indirect deaths from pollution caused by the world's coal-powered stations and the total is a staggering five million or more each year.*

This is a rarely noted and quite astounding fact!

http://news.uns.purdue.edu/x/2007a/070515WoodallHydrogen.html

New process generates hydrogen from aluminum alloy to run engines, fuel-cells

PURDUE UNIVERSITY, WEST LAFAYETTE, Ind. May 15, 2007

Used with permission

A Purdue University engineer has developed a method that uses an aluminum alloy to extract hydrogen from water for running fuel-cells or internal combustion engines, and the technique could be used to replace gasoline.

The method makes it unnecessary to store or transport hydrogen—two major challenges in creating a hydrogen economy, said Jerry Woodall, a distinguished professor of electrical and computer engineering at Purdue who invented the process.

"The hydrogen is generated on demand, so you only produce as much as you need when you need it," said Woodall, who presented research findings detailing how the system works during a recent energy symposium at Purdue.

The technology could be used to drive small internal combustion engines in various applications, including portable emergency generators, lawn mowers and chain saws. The process could, in theory, also be used to replace gasoline for cars and trucks, he said.

Hydrogen is generated spontaneously when water is added to pellets of the alloy, which is made of aluminum and a metal called gallium. The researchers have shown how hydrogen is produced when water is added to a small tank containing the pellets. Hydrogen produced in such a system could be fed directly to an engine, such as those on lawn mowers.

"When water is added to the pellets, the aluminum in the solid alloy reacts because it has a strong attraction to the oxygen in the water," Woodall said.

This reaction splits the oxygen and hydrogen contained in water, releasing hydrogen in the process.

The gallium is critical to the process because it hinders the formation of a skin normally created on aluminum's surface after oxidation. This skin usually prevents oxygen from reacting with aluminum, acting as a barrier. Preventing the skin's formation allows the reaction to continue until all of the aluminum is used.

The Purdue Research Foundation holds title to the primary patent, which has been filed with the U.S. Patent and Trademark Office and is pending. An Indiana startup company, AlGalCo LLC., has received a license for the exclusive right to commercialize the process.

The research has been supported by the Energy Center at Purdue's Discovery Park, the university's hub for interdisciplinary research.

"This is exactly the kind of project that suits Discovery Park. It's exciting science that has great potential to be commercialized," said Jay Gore, associate dean of engineering for research, the Energy Center's interim director and the Vincent P. Reilly Professor of Mechanical Engineering.

The research team is made up of electrical, mechanical, chemical and aeronautical engineers, including doctoral students.

Woodall discovered that liquid alloys of aluminum and gallium spontaneously produce hydrogen if mixed with water while he was working as a researcher in the semiconductor industry in 1967. The research, which focused on developing new semiconductors for computers and electronics, led to advances in optical-fiber communications and light-emitting diodes, making them practical for everything from DVD players to automotive dashboard displays. That work also led to development of advanced transistors for cell phones and components in solar cells powering space modules like those used on the Mars rover, earning Woodall the 2001 National Medal of Technology from President George W. Bush.

"I was cleaning a crucible containing liquid alloys of gallium and aluminum," Woodall said. "When I added water to this alloy—talk about a discovery—there was a violent poof. I went to my office and worked out the reaction in a couple of hours to figure out what had happened. When aluminum atoms in the liquid alloy come into contact with water, they react, splitting the water and producing hydrogen and aluminum oxide.

"Gallium is critical because it melts at low temperature and readily dissolves aluminum, and it renders the aluminum in the solid pellets reactive with water. This was a totally surprising discovery, since it is well known that pure solid aluminum does not readily react with water." The waste products are gallium and aluminum oxide, also called alumina. Combusting hydrogen in an engine produces only water as waste.

"No toxic fumes are produced," Woodall said. "It's important to note that the gallium doesn't react, so it doesn't get used up and can be recycled over and over again. The reason this is so important is because gallium is currently a lot more expensive than aluminum. Hopefully, if this process is widely adopted, the gallium industry will respond by producing large quantities of the low grade gallium required for our process. Currently, nearly all gallium is of high purity and used almost exclusively by the semiconductor industry." Woodall said, "Because the technology makes it possible to use hydrogen instead of gasoline to run internal combustion engines it could be used for cars and trucks.

In order for the technology to be economically competitive with gasoline, however, the cost of recycling aluminum oxide must be reduced," he said.

"Right now it costs more than $1 a pound to buy aluminum, and, at that price, you can't deliver a product at the equivalent of $3 per gallon of gasoline," Woodall said.

"However, the cost of aluminum could be reduced by recycling it from the alumina using a process called fused salt electrolysis. The aluminum could be produced at competitive prices if the recycling process were carried out with electricity generated by a nuclear power plant or windmills. Because the electricity would not need to be distributed on the power grid, it would be less costly than power produced by plants connected to the grid, and the generators could be located in remote locations, which would be particularly important for a nuclear reactor to ease political and social concerns," Woodall said.

"The cost of making on-site electricity is much lower if you don't have to distribute it," Woodall said.

"The approach could enable the United States to replace gasoline for transportation purposes, reducing pollution and the nation's dependence on foreign oil. If hydrogen fuel-cells are perfected for cars and trucks in the future, the same hydrogen-producing method could be used to power them," he said.

"We call this the aluminum-enabling hydrogen economy," Woodall said. "It's a simple matter to convert ordinary internal combustion engines to run on hydrogen. All you have to do is replace the gasoline fuel injector with a hydrogen injector. Even at the current cost of aluminum the method would be economically competitive with gasoline if the hydrogen were used to run future fuel-cells.

"Using pure hydrogen, fuel-cell systems run at an overall efficiency of 75 percent, compared with 40 percent using hydrogen extracted from fossil fuels and with 25 percent for internal combustion engines," Woodall said. "Therefore, when and if fuel-cells become economically viable, our method would compete with gasoline at $3 per gallon even if aluminum costs more than a dollar per pound. The hydrogen-generating technology paired with advanced fuel-cells also represents a potential future method for replacing lead-acid batteries in applications such as golf carts, electric wheel chairs and hybrid cars," he said.

The technology underscores aluminum's value for energy production. "Most people don't realize how energy intensive aluminum is," Woodall said. "For every pound of aluminum you get more than two kilowatt hours of energy in the form of hydrogen combustion and more than two kilowatt hours of heat from the reaction of aluminum with water. A mid-size car with a full tank of aluminum-gallium pellets, which amounts to about 350 pounds of aluminum, could take a 350-mile trip and it would cost $60, assuming the alumina is converted back to aluminum on-site at a nuclear power plant.

"How does this compare with conventional technology? Well, if I put gasoline in a tank, I get six kilowatt hours per pound, or about two and a half times the energy than I get for a pound of aluminum. So I need about two and a half times the weight of aluminum to get the same energy output, but I eliminate gasoline entirely, and I am using a resource that is cheap and abundant in the United States. If only the energy of the generated hydrogen is used, then the aluminum-gallium alloy would require about the same space as a tank of gasoline, so no extra room would be needed, and the added weight would be the equivalent of an extra passenger, albeit a quite large extra passenger."

The concept could eliminate major hurdles related to developing a hydrogen economy. Replacing gasoline with hydrogen for transportation purposes would require the production of large quantities of hydrogen, and the hydrogen gas would then have to be transported to filling stations. Transporting hydrogen is expensive because it is a *nonideal gas,* meaning storage tanks contain less hydrogen than other gases.

"If I can economically make hydrogen on demand, however, I don't have to store and transport it, which solves a significant problem," Woodall said.

Writer: Emil Venere, (765) 494-4709, **venere@purdue.edu**

Sources: Jerry M. Woodall, (765) 494-3479

woodall@dynamo.ecn.purdue.edu

Jay Gore, (765) 494-2122, **gore@purdue.edu**

AlGalCo LLC contact, **Ksquaredtrdgco@aol.com**

Purdue News Service: (765) 494-2096; **purduenews@purdue.edu**

Note: A video describing how the hydrogen-producing technology works is available online at: **http://hydrogen.ecn.purdue.edu**

Web Site Photo Caption:

Purdue researchers demonstrate their method for producing hydrogen by adding water to an alloy of aluminum and gallium. The hydrogen could then be used to run an internal combustion engine. The reaction was discovered by Jerry Woodall, center, a distinguished professor of electrical and computer engineering. Charles Allen, holding test tube, and Jeffrey Ziebarth, both doctoral students in the School of Electrical and Computer Engineering, are working with Woodall to perfect the process. (Purdue News Service photo/ David Umberger)

A publication-quality photo is available from the News Service at:

http://news.uns.purdue.edu/images/+2007/woodall-hydrogen3.jpg

http://www.news.wisc.edu/13881

Full article from this Web site, used with permission

Engineers develop higher-energy liquid fuel from sugar

June 20, 2007, by James Beal

Plants absorb carbon dioxide from the air and combine it with water molecules and sunshine to make carbohydrate or sugar. Variations on this process provide fuel for all of life on earth.

Reporting in the June 21 issue of the journal Nature, University of Wisconsin-Madison chemical and biological engineering Professor James Dumesic and his research team describe a two-stage process for turning biomass-derived sugar into 2,5-dimethylfuran (DMF), a liquid transportation fuel with 40 percent more energy density than ethanol.

The prospects of diminishing oil reserves and the threat of global warming caused by releasing otherwise trapped carbon into the atmosphere have researchers searching for a sustainable, carbon-neutral fuel to reduce global reliance on fossil fuels. By chemically engineering sugar through a series of steps involving acid and copper catalysts, salt and butanol as a solvent, UW Madison researchers created a path to just such a fuel.

"Currently, ethanol is the only renewable liquid fuel produced on a large scale," says Dumesic. "But ethanol suffers from several limitations. It has relatively low energy density, evaporates readily, and can become contaminated by absorption of water from the atmosphere. It also requires an energy intensive distillation process to separate the fuel from water." Not only does dimethylfuran have higher energy content, it also addresses other ethanol shortcomings. DMF is not soluble in water and therefore cannot become contaminated by absorbing water from the atmosphere. DMF is stable in storage and, in the evaporation stage of its production, consumes one-third of the energy required to evaporate a solution of ethanol produced by fermentation for biofuel applications.

Dumesic and graduate students Yuriy Roman-Leshkov, Christopher J. Barrett and Zhen Y. Liu developed their new catalytic process for creating DMF by expanding upon earlier work. As reported in the June 30, 2006, issue of the journal Science, Dumesic's team improved the process for making an important chemical intermediate, hydroxymethylfurfural (HMF), from sugar.

Industry uses millions of tons of chemical intermediates, largely sourced from petroleum or natural gas, as the raw material for many modern plastics, drugs and fuels.

The team's method for making HMF and converting it to DMF is a balancing act of chemistry, pressure, temperature, and reactor design. Fructose is initially converted to HMF in water using an acid catalyst in the presence of a low-boiling-point solvent. The solvent extracts HMF from water and carries it to a separate location. Although other researchers had previously converted fructose to HMF, Dumesic's research group made a series of improvements that raised the HMF output and made the HMF easier to extract. For example, the team found that adding salt (NaCl) dramatically improves the extraction of HMF from the reactive water phase and helps suppress the formation of impurities.

In the June 21, 2007, issue of Nature, Dumesic's team describes its process for converting HMF to DMF over a copper-based catalyst. The conversion removes two oxygen atoms from the compound lowering the boiling point, the temperature at which a liquid turns to gas, and making it suitable for use as transportation fuel.

Salt, while improving the production of HMF, presented an obstacle in the production of DMF. It contributed chloride ions that poisoned the conventional copper chromite catalyst. The team instead developed a copper ruthenium catalyst providing chlorine resistance and superior performance.

Dumesic says more research is required before the technology can be commercialized. For example, while its environmental health impact has not been thoroughly tested, the limited information available suggests DMF is similar to other current fuel components.

"There are some challenges that we need to address," says Dumesic, "but this work shows that we can produce a liquid transportation fuel from biomass that has energy density comparable to petrol."

http://www.scientificamerican.com/article.cfm?id=the-dark-horse-in-race-to&page=1

Excerpt—Go to this site to read the entire article.

News, August 28, 2007

Ultracapacitors, the dark horse in the race to power hybrid cars.

Are ultracapacitors the key to making hybrids king of the auto market?

By Larry Greenemeier

Many motorists chuckle smugly after giving their cars a little extra gas to leave a Toyota Prius or some other eco-friendly automobile in the dust. But Toyota and its earth-loving ilk may yet have the last laugh as they cultivate encouraging new advances in ultracapacitor technology that promise to one day put hybrids in the driver's seat.

The most impressive victory so far for the cars, fueled by a combo of electricity and gas, came just weeks ago when an ultracapacitor-equipped Toyota Supra HV-R coupe became the first hybrid to win the 24-hour endurance car race held at Japan's Tokachi International Speedway. The hybrid Supra finished 616 laps of the 5.1-kilometer (roughly three-mile) course—19 more laps than the second-place nonhybrid Nissan Fairlady Z. "The Toyota that won was able to deliver energy more quickly, accelerate faster, and use braking generation more efficiently," says Kevin Mak, an analyst with research and consulting firm Strategy Analytics and author of a recent study that explores the potential for ultracapacitors to complement and possibly even replace batteries in hybrid vehicles. "The days of the large hybrid vehicle battery pack may be numbered," he adds.

http://www.sciencenews.org/articles/20070505/fob7.asp

Excerpt—Go to this site to read the entire article.

Week of May 5, 2007; Vol. 171, No. 18, p. 278

Ethanol may not be good for the environment. Not-so-clear alternative: in its air-quality effects, ethanol fuel is similar to gasoline

Aimee Cunningham

Switching the nation's vehicles from gasoline to mostly ethanol will not reduce air pollution, a new study finds. The work joins other evidence questioning the benefits of ethanol fuel.

Mark Z. Jacobson, an atmospheric scientist at Stanford University, created a model that takes into account how the chemicals emitted in car exhaust transform through reactions in the atmosphere. He combined the resulting chemical profile with population and health-effects data to determine the risks associated with each of the fuels.

Jacobson looked at emissions from E85, the 85 percent ethanol, 15 percent gasoline blend considered a potentially large-scale replacement for gasoline. He examined a scenario based on expected emissions in 2020, the first year that he says that E85-ready cars might dominate the roads.

http://www.scientificamerican.com/search/index.cfm?q=GM+Volt&submit.x=27&submit.y=7&submit=submit

Excerpt—Go to this site to read the entire article.

GM Announces New Batteries for Chevrolet Volt Plugin, Hybrid

By Jui Chakravorty

TRAVERSE CITY, Michigan (Reuters), General Motors Corp. will begin road testing its Chevrolet Volt plugin hybrid in the spring of next year and remains on track to produce the rechargeable car by late 2010, a senior executive said on Thursday.

As the race to bring a mass-market, rechargeable electric vehicle to the market heats up, GM's global product chief Bob Lutz said he expects to have next-generation lithium-ion battery packs ready for the vehicles by October this year.

"We should have the battery packs by October," he said, speaking to reporters on the sidelines of an industry conference. "We'll have some on the road for testing next spring, and we should have the Volt in production by the end of 2010." GM is the only automaker to have provided a time line on the production of a plugin hybrid vehicle, even though other companies, such as Ford Motor Co. and Toyota Motor Corp. are working on similar technology.

Author's note: *The last paragraph is not factually correct. For some reason Scientific American's writers seem to forget or ignore small and start-up auto companies like AC Propulsion (eBox), Phoenix Motors (Phoenix SUT & SUV) and Tesla Motors (Tesla Roadster). Not only have these new companies provided a time line, but—as this is written—they are delivering vehicles to customers. Not too many years ago and under similar circumstances, Scientific American would probably have ignored companies like Toyota, Nissan, Honda, Hyundai and numerous others as well. My, how times change—even in the science business.*

This section has information about EVs (Electric Vehicles) and PEVs (Plugin Electric Vehicles) and web sites with more information.

The eBox and the tzero PEVs **http://acpropulsion.com/**

http://www.greencar.com/articles/phoenix-electric-sports-utility-truck.php

The Phoenix SUT **http://www.phoenixmotorcars.com/**

More info about EVs.

http://www.businessweek.com/autos/ Search for hybrids and EVs

Also visit: **http://www.ev1.org/** for some amazing food for thought about *Who Killed the Electric Vehicle, GM and Chevron?* Links to many web sites with information on EVs and PEVs and why successful Electric Vehicles were manufactured by GM, Toyota and Honda and then crushed rather than being sold. Some of this information is scary.

A few comments about this situation: GM claims the EV1 cost them *$1 Billion* to produce.

The Tesla EV sports car only cost $40 Million to develop and is now selling to the public. There's a waiting list.

AC Propulsion spent even less than Tesla to develop the eBox which is now selling to the public. There's a waiting list.

Phoenix motors also spent even less than Tesla to develop their SUED and SUV which are now selling to the public. They also have a waiting list.

GM was so wasteful of money that it refused to sell even the last 78 EV1 to recoup $2M of its *investment*.

Toyota, working to meet the Zero Emission Vehicle mandate, set up a production line in 1997 for the *large-format* EV-95 batteries needed for their Toyota RAV4-EV. These EV-95 NiMH batteries, after years of research, were perfected for EVs.

What happened to this production line, and where did those batteries go?

Description of the EV-95 batteries:

Deep Cycle, no memory effect; High energy output for acceleration; Long lifetime, longer than the life of the car—even a Toyota.

Toyota's EV-95 batteries are still running Toyota RAV4-EV cars more than 20,000 miles per year, and for more than 100,000 miles so far. But no more EV-95 batteries can be made, after Chevron sued Toyota. One ponders, "What is the real reason behind Chevron's reluctance to license its EV-95 battery technology for use in electric vehicles?"

Comparison 2007 Bentley ARNAGE with the 2007 Tesla plugin EV

	2007 Bentley ARNAGE	**2007 Tesla plugin EV**
Price	$271,000	$92,000 plus tax
BHP	500 at 3200RPM	EHP 200
Weight	5687 lbs.	3,000 lbs. or less
Acceleration	0 to 60, 5.2 Seconds	0 to 60, 3.8 seconds
Fuel use	11 mpg	125 mpg Equiv
Power plant	6.75 liter OHV V8	DC Motor, 2 sp trans
Top speed	179 mph	130 mph (regulated)
Pollution:	2 lbs. CO_2 /mi system,	None if powered by home solar energy or other nonpolluting source
Range	440 miles on 40 GG (GG = gal gasoline)	250-300 mi/charge equivalent of 2 GG
Depn/mile	$ 1.35	$ 0.46
Op Cost/mile	$ 0.73	$ 0.08
	(20k/year)—included: energy, maintenance and tires	

Comparison of the ACP eBox with the Toyota Prius

	AC Propulsion eBox	Toyota Prius
Price	$70,000	$20,000-$25,000
BHP	140 hp, 125 kW	76 hp, 67 kW
Weight	2,970 lbs	2,890 lbs
Acceleration	0 to 60, 7 seconds	0 to 60, 9 seconds
Fuel use	25 mpg	Equiv 45 mpg
Power plant	125 kW AC motor	76 hp gas, 67 kW AC
Top speed	95 mph	90 mph
Pollution:	None if powered by nonpolluting source	0.4 lbs. CO_2 per mile
Range	120-150 mi/charge equivalent of 1GG (GG = gal gasoline)	300-350 miles
Depn/mile	$ 0.35	$ 0.12
Op Cost/mi	$ 0.08	$ 0.11

(20k/year)—included:
energy, maintenance and tires

eBox Charge Rate:
30 min for 20 to 50 miles
Full Charge 2 hrs (fast),
5 hrs (normal)

Comparison of the Phoenix SUED with Toyota 4 Dr pickup

	Phoenix SUED	**Toyota 4Dr Pickup**
Price	$50,000	$17,000-$21,000
BHP	100 Kw, 76 Hp	159 hp
Weight	2,970 lbs	5,400 lbs
Acceleration	0 to 60, 10 seconds	0 to 60, 9 seconds
Fuel use	100 mpg Equiv	21 mpg
Power plant	125 Kw AC motor	2.7L I-4
Top speed	95 mph	85 mph
Pollution:	None if powered by home solar system or other nonpolluting source	1.0 lbs. CO_2 per mile
Range	120 mi/charge 250 miles in future equivalent of 1GG	300-350 miles
Depn/mile	$ 0.25	$ 0.12
Op Cost/mile	$ 0.09	$ 0.25

(20k /year) included:
energy, maintenance and tires

Phoenix SUED

Charge Rate: 10 min for 95% external charger, 6 hrs for plug in to 220vac outlet

The break-even mileage for operating costs of Phoenix and Toyota trucks is between 200,000 miles and 240,000 miles. Should the Phoenix enjoy sales growth to substantial numbers, the selling price and thus depreciation will be lowered substantially. If the experience with the GM and Toyota EVs hold true, maintenance costs, reliability and vehicle life will be much better than with gas or diesel pickups.

The Phoenix SUT begins life as a Sangyong Actyon Sports SUED. While Sangyong is Korea's fourth largest automaker, its products are not presently sold in this country. Because the Actyon is unknown in America it presents the look and feel of an all-new, designed-from-the-ground-up, dedicated electric vehicle model since no conventionally-powered counterparts are ever seen on American highways.

Zero emission, all-battery power, NanoSafe production battery pack system provides 100+ miles per charge, speeds of 95 m.p.h.[while] carrying five passengers and full payload.

High torque: 0 to 60 m.p.h. in 10 seconds Long battery pack life: 250,000 miles/12+ years Off board charger: 10 minutes to recharge to 95% capacity Onboard charger: 6 hours to recharge from 220V outlet. To contact Phoenix Motors by email, visit

http://www.phoenixmotorcars.com/

For the top ten electric vehicles of 2007 go to:

http://www.autobloggreen.com/2007/02/07/the-top-ten-electric-vehicles-you-can-buy-today-for-the-most-pa/

H

I

J

www.ingramcontent.com/pod-product-compliance
Lightning Source LLC
Chambersburg PA
CBHW081227090426
42738CB00016B/3210